Analog Circuits and Signal Processing

The Analog Circuits and Signal Processing book series, formerly known as the Kluwer International Series in Engineering and Computer Science, is a high level academic and professional series publishing research on the design and applications of analog integrated circuits and signal processing circuits and systems. Typically per year we publish between 5-15 research monographs, professional books, handbooks, edited volumes and textbooks with worldwide distribution to engineers, researchers, educators, and libraries.

The book series promotes and expedites the dissemination of new research results and tutorial views in the analog field. There is an exciting and large volume of research activity in the field worldwide. Researchers are striving to bridge the gap between classical analog work and recent advances in very large scale integration (VLSI) technologies with improved analog capabilities. Analog VLSI has been recognized as a major technology for future information processing. Analog work is showing signs of dramatic changes with emphasis on interdisciplinary research efforts combining device/circuit/technology issues. Consequently, new design concepts, strategies and design tools are being unveiled.

Topics of interest include:

Analog Interface Circuits and Systems;

Data converters;

Active-RC, switched-capacitor and continuous-time integrated filters;

Mixed analog/digital VLSI;

Simulation and modeling, mixed-mode simulation;

Analog nonlinear and computational circuits and signal processing;

Analog Artificial Neural Networks/Artificial Intelligence;

Current-mode Signal Processing; Computer-Aided Design (CAD) tools;

Analog Design in emerging technologies (Scalable CMOS, BiCMOS, GaAs, heterojunction and floating gate technologies, etc.);

Analog Design for Test;

Integrated sensors and actuators; Analog Design Automation/Knowledge-based Systems; Analog VLSI cell libraries; Analog product development; RF Front ends, Wireless communications and Microwave Circuits;

Analog behavioral modeling, Analog HDL.

More information about this series at http://www.springer.com/series/7381

Mohammad Alhawari • Baker Mohammad
Hani Saleh • Mohammed Ismail

Energy Harvesting for Self-Powered Wearable Devices

Springer

Mohammad Alhawari
Department of Electronic
and Computer Engineering
Khalifa University of Science
and Technology (KU)
Abu Dhabi, United Arab Emirates

Baker Mohammad
Department of Electronic
and Computer Engineering
Khalifa University of Science
and Technology (KU)
Abu Dhabi, United Arab Emirates

Hani Saleh
Department of Electronic
and Computer Engineering
Khalifa University of Science
and Technology (KU)
Abu Dhabi, United Arab Emirates

Mohammed Ismail
Department of Electrical
and Computer Engineering
Khalifa University of Science
and Technology (KU)
Abu Dhabi, United Arab Emirates

ISSN 1872-082X ISSN 2197-1854 (electronic)
Analog Circuits and Signal Processing
ISBN 978-3-319-62577-5 ISBN 978-3-319-62578-2 (eBook)
DOI 10.1007/978-3-319-62578-2

Library of Congress Control Number: 2017947690

Printed on acid-free paper

This Springer imprint is published by Springer Nature
The registered company is Springer International Publishing AG
The registered company address is: Gewerbestrasse 11, 6330 Cham, Switzerland

Mohammad Alhawari would like to dedicate this work to his parents, wife, and family.

Baker Mohammad would like to dedicate this work to the memory of his parents.

Hani Saleh would like to dedicate this work to the memory of his parents. May Allah shower them with his mercy.

M. Ismail would like to dedicate this work to the memory of his parents, Ismail A. Elnaggar and Sameha Elsharkawy.

Preface

The ever increasing demand for computing and communications especially for wearable devices, coupled with some of the unique applications that require non-interruptible operation, dictates the need for innovation in energy harvesting and power management. Further, the advancements in circuit design, systems and communication coupled with semiconductor technology scaling have enabled ultra-low power systems in the range of microwatt to nanowatt. This development along with the necessity for near perpetual operations, especially for wearable devices, low cost and small size, brings new challenges to energy sources and power managements. In addition, the single digit improvement in battery technology has resulted in increasing the focus on energy harvesting as the main power source.

The usage of wearable devices in healthcare has led to a proliferation of research toward smaller, smarter, connected systems operating at low power consumption. These devices have attracted huge interest to monitor different vital signs over long time such as heart rate, oxygen level, breath rate, and glucose rate. In addition, having this information monitored continuously could help doctors to track the health status of their patients without being physically in the hospital. This reduces the effort, especially for elderly patients, of frequent visits to the hospitals. In healthcare, it is critical to have long time operation due to the fact that data interruption is not preferred especially when monitoring vital signs. Further, the inconvenience of recharging/replacing batteries has pushed to find better solutions to extend their lifetime. Energy harvesting is an emerging topic that can be utilized to solve the issue of a short lifetime for wearable applications. Although energy harvesting has been known for decades, it is now becoming more feasible to consider for applications in wearable devices and Internet-of-Things (IoT) since the power consumption requirements of such devices are normally low enough and can, in fact, be supported by energy harvesting sources. In addition, state-of-the-art harvesters have become small, efficient, and flexible which allow them to be integrated into a small form factor that fit the wearables. Furthermore, energy harvesters could be used directly to power devices without the existence of a battery. The battery-free operation enables new classes of applications in medical devices, environmental sensors, hard to reach places, smart buildings, and implantables.

This book provides breadth and depth coverage of energy harvesting techniques with a focus on wearable devices. It reflects the cutting-edge technology in designing a complete energy harvesting systems starting from the harvester model and analysis toward powering different systems. This includes deep details on circuit topologies and control circuits that are necessary to achieve high efficient multi-source energy harvesting system. Although the focus of this book is on energy harvesting in general, however, thermal harvesting have received a major concentration due to its global availability in wearable applications.

The book is structured as follows. The first chapter explains the latest trend in the wearable application within the era of the Internet-of-Things and the need for energy harvesting. Further, a comparison between using batteries and capacitors is carried out to power wearable devices. In addition, state-of-the-art low power devices that are reported in the literature are compared.

Chapter 2 explores common energy harvesting sources and provides details about their electrical model along with some measurements examples. In addition, different power conversion circuits are presented that are commonly used in energy harvesting applications. This includes linear dropout regulator, switch capacitor circuits, and inductor-based converters.

Chapter 3 supplies the reader with the main interface circuits and control techniques for energy harvesting in general focusing on human body thermal harvesting. This includes the design of high gain inductor-based boost converter with the associated control techniques.

In Chap. 4, zero crossing switching methods for thermal harvesting are explained and compared in regard to design, complexity, and efficiency. Also, a new technique is proposed with measurement results that enhances the efficiency as well as the dynamic of the inductor boost converter.

Polarity circuits for thermoelectric generators are demonstrated in Chap. 5 and a novel technique is proposed and supported by measurement results. This technique is simple yet effective, fully integrated on-chip, all digital with low overhead.

Finally, the design of energy combiner with power manager to form a complete energy harvesting system is explained in Chap. 6. In addition, the design of an efficient energy combiner is proposed with power manager circuit. Furthermore, a sleep mode operation for low power processor is explained which is necessary when the input energy is variable.

This book could be used as a reference for design engineers, practitioners, scientists, and marketing managers in the semiconductor industry developing integrated self-powered platform system-on-chip solutions for wearable devices and the Internet-of-Things. It is also highly recommended for graduate students in electrical and computer engineering and physics pursuing research in integrated devices and circuits, energy harvesting, power management as well as overall system integration. This book could also be useful for researchers that are in other disciplines since the material is rich with basic information that familiarizes the reader with the energy harvesting subject.

Acknowledgments

The work in this book has its roots in the doctoral dissertation of the first author. We would like to thank and acknowledge all those who assisted us with the different phases of developing the material that led to this book. We would like to specifically acknowledge our colleagues at the Khalifa Semiconductor Research Center (KSRC) for their help, encouragement, and support, thanks to Temesghen, Yonatan, Dima, Maisam, Lama, Yarjan, and Nourhan Bayasi. Also, thanks to Hadeel, Nasma, Nadeen, and Ayaa for their help in RF energy harvesting.

We also like to acknowledge the support of Mubadala for the funding and the US Semiconductor Research Corporation (SRC) for overseeing the projects of the ACE^{4}S Mubadala-SRC Center of excellence under which this project was completed. We must also acknowledge our industrial liaisons for their suggestions and insights: John Pigott and Mark Schlarmann (from NXP), and Muhammad Khellah and Lilli Huang (from Intel).

The work in this book was part of a complete system-on-chip targeting a platform for wearable healthcare. The authors would like to thank their colleagues responsible for other parts of such a system and acknowledge their unmatched spirit of teamwork. Finally, we would like to acknowledge the help and support of our families and friends and thank them for their patience and understanding.

Abu Dhabi, United Arab Emirates — Mohammad Alhawari
Abu Dhabi, United Arab Emirates — Baker Mohammad
Abu Dhabi, United Arab Emirates — Hani Saleh
Abu Dhabi, United Arab Emirates — Mohammed Ismail

Contents

Chapter 1
Introduction

1.1 Wearable Devices and Battery Technology in IoTs

Internet-of-Things (IoT) has become a growing concept with a potential to affect how we live and how we communicate with the "Livings" and the "Things" around us. IoT is a concept of connecting all the daily use devices into the internet such as mobile phones, refrigerator, computers, doors, gadgets, and cars. These devices not only can be connected to the internet but also can talk with each other. This way, people can talk to their devices, and devices can respond by talking to other devices. For example, a person can speak to his smart car to turn on the air condition before he arrives home. This revolution of IoT has also opened the door to a tremendous number of services that have made the human's life much easier. For example, Fig. 1.1 shows a typical diagram of what IoT sections look like [1]. It consists of smart devices that are connected through a communication channel to the Internet. Then the cloud provides many services to connect different devices together.

Since the IoT paradigm is to connect many "things" together which will enable smarter and more intelligent devices that improve the quality of life. The projection of IoT devices is expected to reach billions in the next decade or so [2]. With billions of devices connected together, many challenges are faced. Security is a big threat in IoT where data needs to be transferred and stored in a secured fashion. Also, the power consumption and the power management of the devices especially for battery-powered devices. To put a perspective on the size of this issue, if we have a trillion IoT devices that run on batteries, then we need 274 million battery change per day [3]. This put a huge pressure on the environmental issues from battery disposal and recycling.

Wearable devices have emerged from the IoT in which they are smart, portable, multi-functional, connected to the internet, and run on batteries. Wearable devices have been used in many application such as mobile phones, sports athletes, and smart glasses. In addition, wearables have also enabled endless possibilities on

M. Alhawari et al., *Energy Harvesting for Self-Powered Wearable Devices*,
Analog Circuits and Signal Processing, DOI 10.1007/978-3-319-62578-2_1

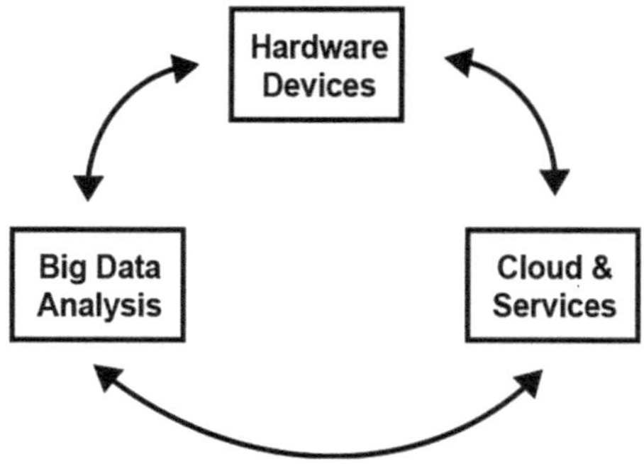

Fig. 1.1 IoT sections that include hardware, data, and cloud

how these devices could be used. For example, new trends in the technology have emerged that will impact how we live such as Virtual Reality (VR) and Artificial Intelligence (AI). Further, the usage of wearable devices in healthcare has led to an enormous research toward smaller, smarter, connected systems operating at low power consumption. These devices have attracted huge interest due to the need for monitoring different vital signs over longtime duration such as heart rate, oxygen level, breath rate, and glucose rate. In addition, having this information monitored continuously could help doctors to track the health status of their patients without being physically in the hospital. This reduces the effort, especially for elderly patients, of frequent visits to the hospitals. For example, a wearable patch placed on the patient's chest could be used to track the heart's status through monitoring the electrocardiography (ECG). The patch could also have a wireless communication to send the data to a mobile phone as well as connected to a server in the hospital for remote monitoring.

The main energy source for most wearable devices is the battery. Hence, battery lifetime has become one of the main issues for perpetual operation in wearable electronics [4]. In healthcare, it is even more critical due to the fact that data interruption is not preferred especially when monitoring vital signs. Further, the inconvenience of recharging/replacing batteries has pushed the need to find better solutions to extend their lifetime. Finally, battery disposal and recycling have a huge environmental impact.

Lithium-ion battery is the most used type in our electronic devices due to its high energy density, low leakage, low memory effect, and long lifetime. However, Li-ion batteries suffer from high cost, high-temperature sensitivity, aging effect, and safety concerns. Further, the improvement rate of Li-ion batteries has been increasing around 5–8% per year for the last three decades. This rate is not even close to the improvement in the semiconductor technology (Moore's law) which is doubling the number of transistors every 18 months. Hence, batteries have been usually the larger, heavier, and most expensive component in wearable devices.

With the semiconductor improvement, wearable devices have become more sophisticated that perform sensing, processing, and wireless connectivity to the Internet in a very small factor. However, the lifetime of such device is usually 1 day or even less since more power is required to perform all the functionalities. As such, the purpose of having the wearability for a device that lasts for a long time is defeated because of the limited battery capacity within the device's form factor.

1.2 Energy Harvesting and Autonomous Systems

Recently, there has been an increase in activities to find solutions for extending the lifetime operation for wearables. Energy harvesting is an emerging topic that can be utilized to recharge/replace batteries in low power application. Although, energy harvesting has been known for decades, however, it is becoming possible to harvest since the power consumption of electronic circuits has become low enough to be powered by energy harvesting sources [5, 6]. Further, recent researches in circuit design have pushed the power consumption of electronic circuit as low as possible. In addition, state-of-the-art harvesters have become small, efficient, flexible which allow these harvesters to be integrated into small form factor.

The usage of energy harvesting has got an interest in the battery-free operation. This means that the device can harvest the energy from the environment and directly power the load in an efficient manner. The battery-free operation enables new classes of applications in medical devices, environmental sensors, hard to reach places, smart buildings, and implantables [7–9]. In battery-free operation, a mean of storage is still needed in order to maintain the voltage. The use of a capacitor instead of a battery has many advantages. Table 1.1 shows a comparison between batteries and capacitors. The main differences are the charging cycles where capacitors can reach millions of cycles. Furthermore, the charging time is much faster for capacitors where seconds are needed for charging. In addition, capacitors could help in reducing the form factor of the devices which are important in many applications. However, the leakage of capacitors is higher than batteries and the capacity is lower. For the leakage issue, it reduces the efficiency in the device. In contrast, the lower capacity might be suitable in some application due to the low power requirement.

The use of only capacitors instead of batteries requires a sleep mode model designed for that device. To explain, consider an energy harvesting source with a μWatt capability that is used to power a device which has a radio that consumes

Table 1.1 Comparison table between batteries and capacitors

Comparison	Conventional batteries	Thin film batteries	Supercaps
Recharge cycles	100s	5–10k	Millions
Self-discharge	Moderate	Negligible	High
Charge time	Hours	Minutes	Sec-minutes
Impedance	Low–high	High	Low
Physical size	Large	Small	Medium
Capacity	0.3–2500 mAH	12–220 μAH	10–100 μAH

Table 1.2 Power available from different energy sources

Source	Source power	Harvested power
Ambient light		
Indoor	$0.1\,mW/cm^2$	$10\,\mu W/cm^2$
Outdoor	$100\,mW/cm^2$	$10\,mW/cm^2$
Vibration/motion		
Human	0.5 m at 1 Hz	
	$1\,m/s^2$ at 50 Hz	$4\,\mu W/cm^2$
Industrial	1 m at 5 Hz	
	$10\,m/s^2$ at 1 kHZ	$100\,\mu W/cm^2$
Thermal energy		
Human	$20\,mW/cm^2$	$30\,\mu W/cm^2$
Industrial	$100\,mW/cm^2$	$1–10\,mW/cm^2$
RF		
GSM base station	$0.3\,\mu W/cm^2$	$0.1\,\mu W/cm^2$

a mWatt range. For such system to operate, the radio module can be only active for a small period of time such that the voltage drop across the storage capacitor should not exceed a certain range based on the system requirements. As such, the system goes to sleep mode again to allow recovering the voltage drop on the storage capacitor.

There are many energy harvesting sources available such as thermal, vibration, radio frequency (RF), and solar energy. Table 1.2 shows the power available from different energy sources [10]. Inspecting of Table 1.2 reveals many key points regarding the power available and power density for each power source. First, the highest power density is from the ambient light where it can be harvested indoor and outdoor as well. However, the availability of the solar limits its application where another energy source should be used during the absence of light. The vibration is considered a good energy source especially in industrial applications where high voltage can be generated with moderate power density. The thermal energy has an interest in many application especially for wearable devices from human body heat. RF energy can be used in places where another source might be partially/not available such as implantable devices inside the human body.

To check if the power generated from energy harvesting sources could run low power devices, we need to explore different parts of these devices. Figure 1.2 displays a general block diagram of a wearable device. It consists of sensors that can be either passive or active. These sensors send analog data to an analog front end (AFE) to be amplified, filtered, and digitized. The Analog to Digital Converter (ADC) provides the digital value to the processor unit. The processor is programmed to process the data and then send it to the radio. Finally, the radio sends the data outside the devices where it can be displayed to the user. The energy system includes all the power circuit to convert the input voltage from an energy source into different voltage levels suitable for each part of the device.

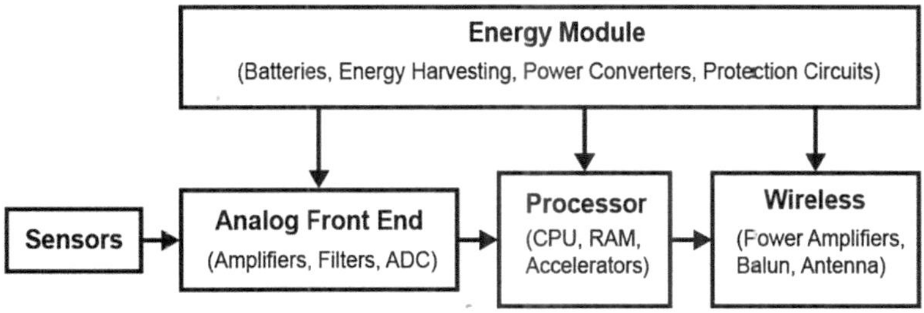

Fig. 1.2 General block diagram of a wearable device

Table 1.3 Power consumption breakdown of state of the art in a wearable device

Component	Power	Comments
Inst. amplifier [Song, ISCAS'13]	1 μW	1.8 V VDD, 250 nV/(Hz)$^{0.5}$
ADC [Yang, EDSSC'15]	212 nW	1 V VDD, 7.4ENOB
16b μ-cont [Kwong ISSCC'08]	2.72 μW	0.5 V VDD, 128 kb SRAM 100 kHz
Radio [CC 2550]	33.6 mW (active)	3 V VDD, 2.4 GHz, −12 dBm Pout

Given the block diagram in Fig. 1.2, Table 1.3 shows examples of a breakdown of the power consumption for different parts in wearable devices reported in the literature. As depicted in the table, the power consumption of the modules is in the range of μW except for the radio. For the radio part, the system could be designed to be active for a small period of time so that the total active energy for the radio is minimized.

Comparing the data from Table 1.3 and various ultra-low power circuits that have been reported in the literature [11–14] with the available power in Table 1.2 reveals that such systems could potentially be powered by energy harvesting and achieve autonomous operation [15].

At present, many state-of-the-art biomedical low power systems are exploring energy harvesting power sources. There are many low power biomedical devices that target longtime monitoring for vital signs of a human body. However, only a few examples will be discussed since there are a large number of solutions in the literature, and doing a detailed comparison is beyond the scope of this book. Google smart lens has been built to detect glucose in tears. The lens integrates the antenna that is used to transmit data as well as receives the power to run the electronics. Further, the lens has all the required components for a complete system which includes storage capacitor, controller, and sensor.

A complete system that comprises a powering coil, a piezo-resistive MEMs pressure sensor, a transmitter antenna, and all the required circuitry in a volume less than 0.5 mm^3 is reported in [16]. The system monitors the pressure in the interior chamber of the eye in order to indicate Glaucoma. The complete system consumes less than 50 μW at 7% TX duty cycle. An implantable device that uses ultrasonic technology to transfer power, transmit, and receive data is reported in [17]. There are two piezoelectric materials that are used for transmitting and receiving. Further, an off-chip capacitor is used as a storage element. In addition, all the required circuits for power management, FSM, PA are integrated on-chip. The overall system has a volume of 30 mm^3 with TX peak power of 177 μW.

In [18], a biomedical system-on-chip (SoC), that is powered by thermal and RF energy harvesting, is fabricated. The SoC integrates all the components for autonomous operation, such as AFE, PMU, biomedical processor, and wireless transmitter. Thermal energy harvesting using a thermoelectric generator (TEG) is utilized to supply the necessary power to all blocks in the SoC. In addition, an RF source is used to initiate startup for the TEG. The total power consumption of the whole system is 19 μW at TX duty cycle of 0.013%. Cheng et al. [19] proposed a wireless body sensor network that is powered by RF energy harvesting. The low voltage supply used (0.56 V) at near-threshold operation helped in reducing the power consumption of the ECG processor to 150 μW which allowed the usage of the RF energy harvesting and without using a battery.

Solar and TEG energy harvesting have been used to power biomedical chips as in [20]. The system supports different applications besides ECG monitoring such as fibrillation detection, fall detection, and activity logging. In addition, the system integrates power management, regulators, and wireless transmitter. The total power consumption of the SoC including the wireless is 7 μW, which enabled the energy harvesting to be used as the power source of the system. MEMs-based wireless energy harvesting has been proposed in [21]. It uses a MEMs-based piezoelectric energy harvesting. The energy harvester is capable of generating a few μW that is enough to power an ECG processor. Other systems in [22–24] provide similar low power biomedical chips that can be powered by energy harvesting.

Overall, renewable energy source has been investigated for many decades, however, they increased in importance recently due to their suitability to power the contemporary low power electronic device. In addition, circuit design techniques played a major role in reducing the power consumption of the devices and extending the lifetime of the batteries. Unlike the battery technology, which could not improve with the CMOS technology with approximately 5% per year [25]. In addition, power management units with their power reduction techniques, such as power and clock gating [11], provided the basic techniques for sleep mode operations to reduce the power consumption of the devices. These low power techniques helped in integrating energy harvesting sources so that batteries can be recharged continuously or even removed to achieve self-powered, free-battery, autonomous operation.

Chapter 2
Energy Harvesting Sources, Models, and Circuits

2.1 Energy Harvesters

There are many renewable energy sources that can be utilized to harvest energy and then use that energy to power electronic devices. The most common sources that can be used for wearable devices are thermoelectric generators, piezoelectric harvester, RF harvesting, and solar. The harvesters of these sources can be made relatively small so that it can be embedded in small form factor. More details about these energy sources are explained in the following subsections.

2.1.1 Thermoelectric Generators

Energy can be harvested from heat using thermoelectric generator (TEG) which is based on the Seebeck effect which states that electricity can be generated from the temperature gradient between two conductors connected together [26]. Although the basic principle of thermoelectric has been known by Thomas Seebeck for a long time, however, recent researches have started to focus on such technique because recent technology has enabled devices, circuits, processors, and SoCs with ultra-low power consumption that can be powered by a TEG.

Figure 2.1a shows the structure of a single *p–n* legs connected electrically in series. In addition, Fig. 2.1b shows a typical TEG device which consists of many *p–n* legs connected electrically in series and thermally in parallel [18]. The basic principle of a TEG is based on the generation of an electron–hole pair in a semiconductor due to a temperature gradient. The buildup of charge on the cold side produces an electrostatic force (an electric potential) which is related to the temperature difference (ΔT) across the TEG using Eq. (2.1), where S is the Seebeck coefficient of the material used [27]. As such, a current will flow if we connect a

M. Alhawari et al., *Energy Harvesting for Self-Powered Wearable Devices*, Analog Circuits and Signal Processing, DOI 10.1007/978-3-319-62578-2_2

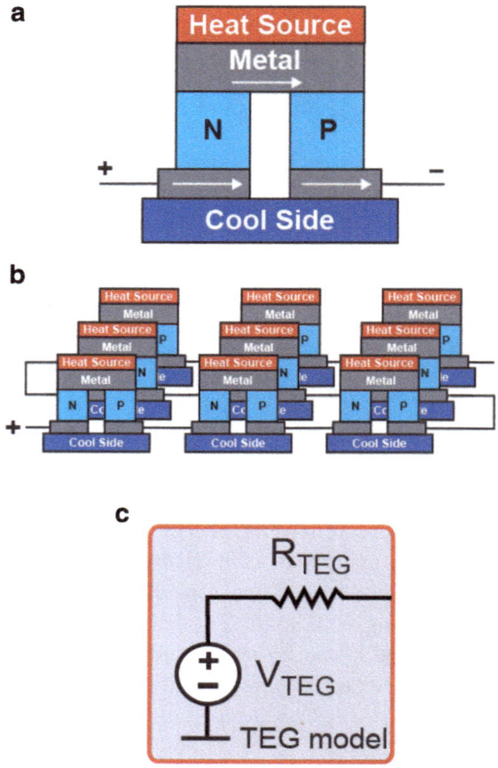

Fig. 2.1 (**a**) *p–n* leg in TEG device. (**b**) TEG device that contains many *p–n* legs connected electrically in series and thermally in parallel. (**c**) Electrical model of the TEG

load across the TEG [10, 28]. Finally, Fig. 2.1c shows the electrical model of the TEG which consists of a voltage source and a series resistance.

$$V_{\text{TEG}} = S \times \Delta T \tag{2.1}$$

Toward the applications side, high efficient TEG systems reflect their viability of extracting electricity from the hot exhaust sources, such as engines in cars, and converting it into a useful energy [29]. In addition, for the special case of implantable devices; using conventional batteries is inconvenient since they require human intervention to replace them [30]. TEG systems are also able to generate power from the sun by using solar energy to create a temperature difference across a TEG material; this is known as the solar thermal energy conversion [29]. Finally, a TEG can be suitable for wearable electronics where human body heat offers a continuous source of energy which could potentially replace batteries.

The figure of merit of a TEG material is shown in Eq. (2.2). where σ and κ are the electric conductivity and thermal conductivity, respectively. Note that when connecting the thermocouples in the TEG electrically in series, the sum of the open circuit voltage of all thermocouples is induced at the TEG terminals.

$$\mathrm{ZT} = \frac{S^2 \times \sigma \times T}{\kappa} \tag{2.2}$$

Since σ and κ are material dependent, Eq. (2.2) provides an intuitive way to enhance the ZT of the device. Although such kind of research requires narrowing the focus down to the chemical and physical properties of the material; it also entails understanding and appropriately modifying its electrical properties [30]. In general, a good thermoelectric material is recognized by its high electrical conductivity and low thermal conductivity. As such, the combination of both will contribute to a high thermoelectric conversion efficiency. For example, increasing the number of thermocouples results in increasing the overall open circuit voltage, however, on the price of the area. Further, decreasing the thermal conductivity, κ, increases ZT which corresponds to the TEG efficiency. This means that the material should not transfer heat across the TEG sides such that the temperature difference is maintained. Hence, an insulation material can be added between the two plates of the TEG to reduce the heat transfer across the TEG. Further, the electrical conductivity, σ, affects the internal resistance of the material, thereby the TEG efficiency and the power available. Finally, emerging new materials, such as Graphene, is investigated to achieve better thermoelectric characteristics [30, 31].

To select a TEG for a certain application, different parameters need to be examined. To explain, two commercial TEG harvesters are considered, namely TEG-I (4 mm × 3 mm × 1 mm) which has 400 Ω series resistance, 150 mV/C Seebeck effect (Micropelt) (50$/piece), and TEG-II (30 mm × 30 mm × 4 mm) which has 5 Ω series resistance, 50 mV/C Seebeck effect (Tellurex) (10$/piece). The system size using TEG-I can be smaller compared to TEG-II at a fixed required power. As such, depending on the application, a TEG can be chosen based on the form factor requirements. In contrast, the open circuit voltage and the maximum current from both TEGs. Even though TEG-I is smaller than TEG-II, however, it can generate higher open circuit output voltage which could be useful in startup applications [32].

2.1.1.1 Human Body Thermal Harvesting

As an example in wearable applications, TEG can be used to harvest the energy from the human body heat [18, 33]. Figure 2.2 shows on-body measurements of $3 \times 3\ \mathrm{cm}^2$ commercial TEG device (G230-0313) from Tellurex [33, 34]. The measured internal resistance of the TEG is 7.6 Ω and the Seebeck coefficient extrapolated from the datasheet is 40 mV/C. The TEG device is mounted on a human wrist as shown

(a)

(b)

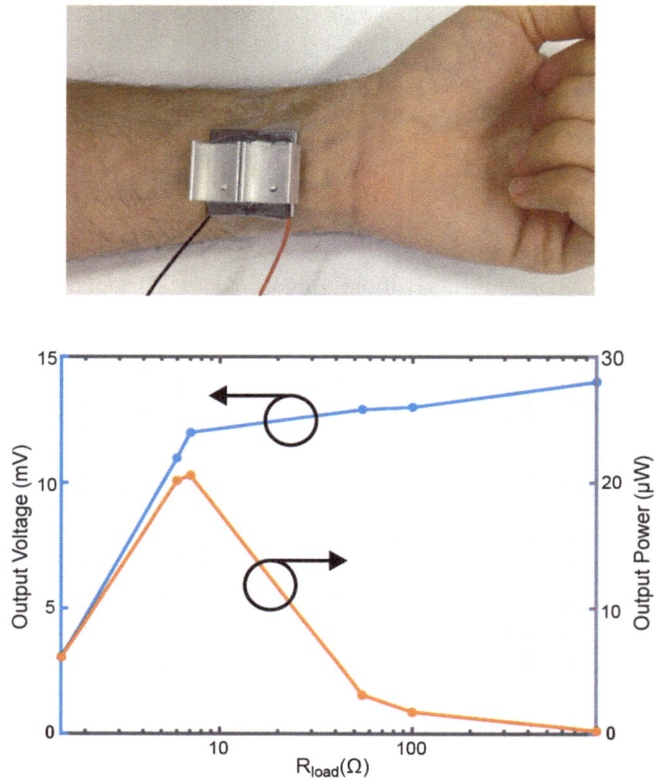

Fig. 2.2 (**a**) Experimental setup of TEG mounted on a human wrist. (**b**) Output voltage and power from human body thermal harvesting

in Fig. 2.2a at room temperature of 22 °C. In addition, as indicated in Fig. 2.2a, a heatsink is mounted on the TEG to improve the heat dissipation and maintain the temperature difference across the TEG.

Figure 2.2b shows the output voltage of the TEG at different resistance values [33]. As depicted in Fig. 2.2b, the maximum power occurs at a load resistance of 7.6 Ω which is equal to 20 μW. This corresponds to a power density of 2.2 μW/cm^2 which is one order of magnitude less than what is reported in [35] (20 μW/cm^2). In [35], a MEMs-based TEG is fabricated with special design in TEG structure as well as the radiator design. In addition, a thermal matching is carefully designed in [35] to maximize the power efficiency. However, due to high fabrication cost, MEMs-based TEG has not been adopted in the industry and BiTe thermopiles are still dominating the TEG mass production. Note that the maximum power limit that can be generated from a wearable TEG placed on a human being is 30 μW/cm^2 at 22 °C ambient temperature and ZT of 1 [36, 37].

The output voltage at the maximum power is approximately 12 mV with an open circuit output voltage of 24 mV. Such low voltage corresponds to approximately 0.5 °C temperature difference across the TEG. This is due to poor thermal matching of the commercial TEG used in this experiment. Finally, the TEG experiment is done in setting condition. However, if the experiment is done while walking, the output power will increase since the TEG will be exposed to the wind during movement and thus more temperature difference [38]. As such, this experiment reports worst case output power from the commercial TEG.

The electrical resistance of the TEG is very important parameter since it limits the maximum power that can be extracted from the TEG. For low-temperature difference, the resistance change is negligible as in the case of human body thermal harvesting [39]. However, in high-temperature application, the resistance change could be substantial which could affect the operation of a certain device. For example, the temperature of a car exhaust could reach larger than 200 °C. Hence, a resistance variation of larger than 40% could be reached.

The use of heat sink for the TEG is essential to get an adequate temperature difference across the device. The heat sink is placed on the cold side of the TEG to reject the heat coming from the hot side. To design an appropriate heat sink for human body thermal harvesting, Eq. (2.3) can be used to calculate the heat sink resistance ($1/K$), where η is the TEG efficiency, ΔT is the temperature difference across the TEG, and P_L is the load output power. For example, for ΔT equals to 3 °C (as in the case of human body thermal harvesting), η of 2%, and P_L of 100 μW, then $1/K$ is 600 °C/W. However, a heat sink with such resistance is not typically found on-the-shelf and needs a custom design. As such, the heat sink needs to be carefully designed to maintain low form factor for wearable devices.

$$\frac{1}{K} = \frac{\Delta T \times \eta}{P_L} \tag{2.3}$$

To extract the maximum power from the TEG, thermal and electrical matching are required. For example, when harvesting from human body heat, the thermal resistance of the skin should match the thermal resistance of the air to ensure maximum heat transfer which is related to the heat sink design discussed above. This puts some design limitations on the TEG-based system such as TEG placement on the human body, TEG fabrication structure, and the radiator design. To first order, the thermal matching is independent of the ambient temperature [40]. Similar to the thermal matching, electrical matching is achieved when the load resistance matches the source resistance (R_{TEG}). This occurs when the output voltage is equal to $V_{\text{TEG}}/2$. The maximum output power is equal to:

$$P_{\text{out,max(TEG)}} = \frac{V_{\text{TEG}}^2}{4R_{\text{TEG}}} \tag{2.4}$$

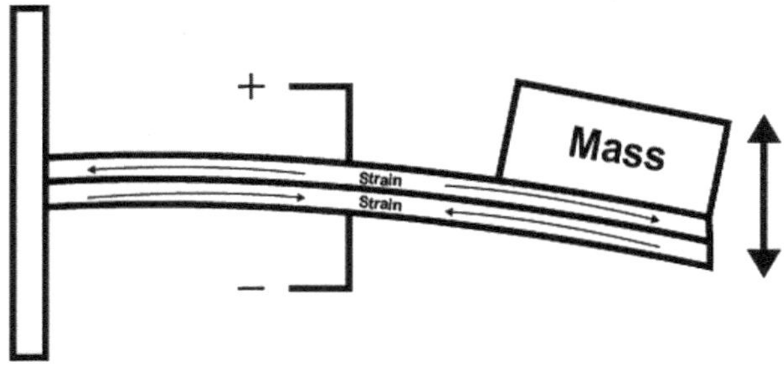

Fig. 2.3 Structure of a piezoelectric harvester

2.1.2 *Piezoelectric Harvesters*

The principle operation of a piezoelectric (PE) device is based on the piezoelectric effect [41]. As shown in Fig. 2.3, when a force is applied to a PE element, a mechanical strain is developed such that the material exhibits a change in its polarization. As such, the positive and negative electric charges accumulate across the PE material. This change in charge distribution develops an electric field, thereby a potential difference. The strength of the potential difference depends on the applied force, the frequency of oscillation, and geometry of the device [42].

The PE harvester model is originally developed from the generic mechanical-to-electrical model [43]. The system associated with this model is the linear spring-mass-damper system. When the system is excited with a force, a simple harmonic motion is generated. Due to damping forces components, such as friction force and drag force, the harmonic motion does not last forever and stops after some time. During the harmonic motion, the mechanical power is transferred into an electrical power. If the device is operated at the resonance condition, the available output power is maximum and depends on the mass of the system (m), the electrical damping ratio (ζ_e), the mechanical damping ratio (ζ_m), the natural (resonance) angular frequency (ω_n), and the input acceleration (a_{in}), as shown in Eq. (2.5).

$$P_{\text{out,max(PE)}} = \frac{m\zeta_e a_{\text{in}}^2}{4\omega_n(\zeta_e + \zeta_m)^2} \tag{2.5}$$

Besides, tip mass affects how rapidly the PE beam oscillations decay from one bounce to the next, i.e., it affects ζ_e and ζ_m. Increasing the tip mass forces the beam oscillations to slowly decay in amplitude toward zero. Therefore, increasing m decreases ζ_e and ζ_m according to Eq. (2.6), where $b_{e,m}$ is the induced damping coefficient [43].

$$\zeta_{e,m} = \frac{b_{e,m}}{2m\omega_n} \tag{2.6}$$

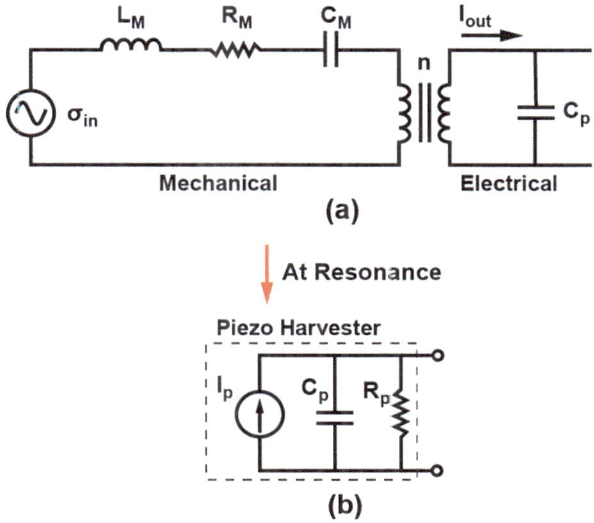

Fig. 2.4 Equivalent circuit of the PE harvester at (**a**) non-resonance condition and (**b**) at resonance [44]

The generic mechanical-to-electrical model is modified so that it is more specific to the PE harvester [44]. The equivalent circuit of the PE harvester is represented as a mechanical spring-mass system coupled to an electrical domain. Figure 2.4a shows the mechanical side and the electrical side of the PE harvester equivalent circuit. The mechanical side has three passive elements that are serially connected: an inductor (L_M) that represents the mechanical mass, a capacitor (C_M) that represents the mechanical stiffness, and a resistor (R_M) that represents the mechanical losses. The electrical side, on the other hand, has a capacitor (C_P) that represents the plate capacitance of the PE material. The mechanical and the electrical domains are coupled by a transformer that converts strain into current [27]. When the PE harvester is excited by sinusoidal vibrations at its resonance frequency, the PE element can be modeled as a sinusoidal current source (I_P) in parallel with a parasitic capacitance (C_P) and a parasitic resistance (R_P) as shown in Fig. 2.4b [27]. This model represents most PE harvesters that are based on cantilever design [45]. The disadvantage of this type of harvesters is that the energy conversion efficiency of the harvester drops substantially if the driving frequency does not match the resonance frequency [45]. Furthermore, C_P and R_P represent two sources of power losses; therefore, the maximum power in the circuit is limited.

Examination of Fig. 2.4b reveals that the transfer function of the PE harvester at resonance can be represented as [46]:

$$\frac{V(s)}{q(s)} = \frac{sR_p}{1 + sR_pC_p} \tag{2.7}$$

where $I(s) = q(s)s$ is the generated current from the PE harvester. The transfer function looks similar to that of a high-pass filter (HPF) and the cutoff frequency (f_c) is

$$f_c = \frac{1}{2\pi R_P C_P} \tag{2.8}$$

2.1.2.1 Human Body Movement Characterization

To characterize the energy generated from human body movement, an off-the-shelf PE harvester (V22BL), named as Volture, is used from MIDE [46]. The Volture is based on the cantilever design and contains two PE harvesters fabricated on the same beam. These PE harvesters can be connected either in series or in parallel.

Connecting the two PE harvesters in series leads to double the open circuit output voltage (V_P) and half the total equivalent parasitic capacitances (C_P) compared to a single PE harvester [46]. In addition, the value of the output current (I_{out}) of two serially connected PE harvesters is the same compared to a single PE harvester [46]. In contrast, connecting the two PE harvesters in parallel leads to same V_P, double C_P, and double I_{out}. The natural frequency (f_n) and the output power ($P_{\text{out(PE)}}$) of the Volture for both cases are the same. The series connection is preferable for applications with low acceleration values [47]. As such, the series connection is chosen for all the experiments in this section. Moreover, C_P of a single Volture is measured using a multimeter. The capacitance for each Volture is approximately 9 nF with a total effective capacitance in the series configuration of 4.6 nF.

The resonance frequency is measured using a simple flick experiment. This is done by flicking the tip of the Volture so that the beam "rings out" as shown in Fig. 2.5. As depicted in the figure, f_n is tuned to 25 Hz. This frequency should be further decreased to match the frequency of human movement in order to maximize the output power. The frequency of human movement is less than 2 Hz. Besides, [46] reported the natural frequency of the PE harvester as 110 Hz instead of 25 Hz. The reason is that [46] measured the natural frequency of the PE harvester by clamping it, while the harvester is not clamped in this setup.

Figure 2.6 shows the measured f_n versus the added tip mass. As depicted in the figure, f_n decreases with increasing the tip mass using the spring-mass system equation:

$$\omega_n = \sqrt{\frac{k}{m}} \tag{2.9}$$

where $\omega_n = 2\pi f_n$ and k is the spring stiffness. Small magnets are used as tip masses and glued together on the tip of the Volture using an adhesive. The number of the magnets is adjusted so that the added tip mass is 3.14 g. This particular mass weight is chosen so that the Volture does not bend too much during vibration which could break the beam. Nonetheless, the mass is essential in increasing the deflection of the

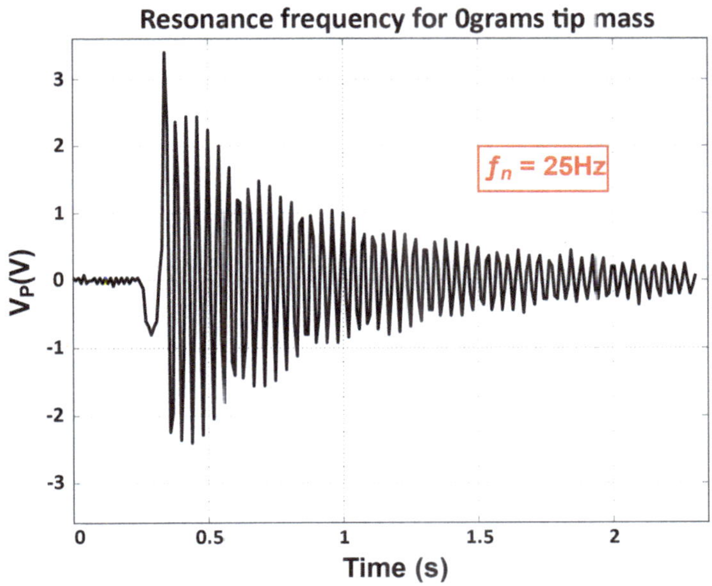

Fig. 2.5 Output voltage of the PE harvester during the flicking experiment at 0 g tip mass

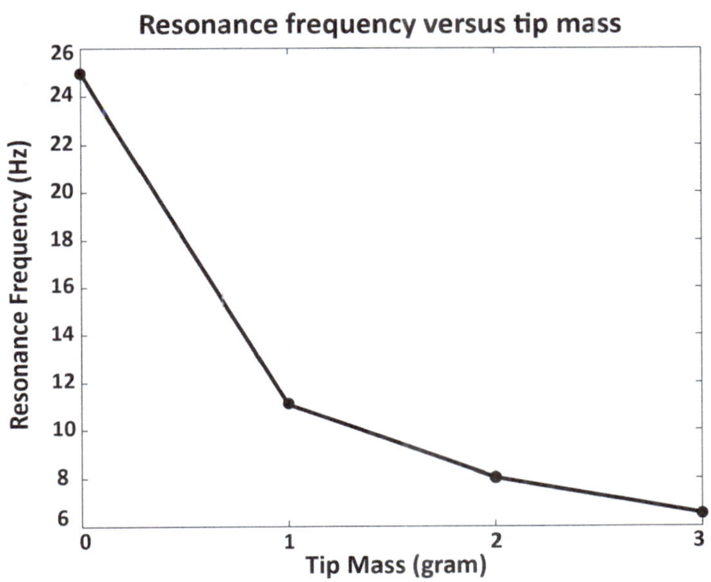

Fig. 2.6 Resonance frequency of the Volture versus tip mass

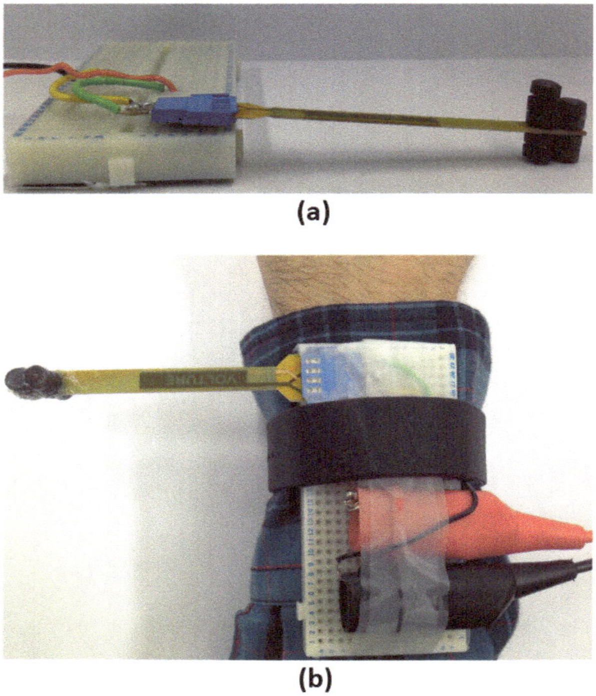

Fig. 2.7 (**a**) Side view of the Volture with the added tip mass and (**b**) the taped Volture mounted on a breadboard and fastened on the experimenter's wrist

PE harvester and decreasing its natural frequency so that it is as close as possible to motion frequency. Although the current setup for the harvester and the tip mass is not practical, however, the harvester could be developed, designed, and fabricated to have the tip mass as part of the harvester itself.

Five main experiments are performed to characterize human body movement using the PE harvester. For the experimental setup, Fig. 2.7a shows the Volture mounted on a breadboard with 3.14 g tip mass. Further, Fig. 2.7b shows the Volture fastened around the experimenter's wrist. Although the harvester seems to be not practical, however, the experiments provide an important data about human body activity using the PE harvester, such as motion frequency and acceleration, harvester output voltage, and suitable tip mass for frequency matching. These data provide a useful information for different parameters necessary for designing more practical PE device.

The first experiment is done to measure the frequency and the acceleration amplitude of human body vibrations using the Slam Stick data logger [48]. The second experiment focused on studying the effect of tip mass on V_P. Initially, the experimenter walked on a treadmill while the PE harvester is attached to his wrist with 0 g tip mass. Then, the value of the tip mass is increased gradually and V_P is monitored using an oscilloscope. The third experiment is done to measure V_P for

3.14 g tip mass at different human speeds: 1.8, 2.5, 3.8, 4.4, 5, 6, and 7 mph. In the fourth experiment, the Volture is connected to a Voltage Doubler (VD) rectifier, and the VD is connected to an output capacitor (C_{rect}) in order to measure the output energy at different speeds. Finally, the fifth experiment is done to measure the output power at different speeds when the PE harvester is connected to a VD. More details about each experiment along with the obtained results will be discussed in the following subsections.

Experiment (1): Acceleration and Frequency of Human Activity To characterize the vibration and frequency profile of human activity, a data logger (VR001) from MIDE is used [48]. The data logger contains 3-axis accelerometer so that the time domain acceleration, as well as the frequency of movement, can be measured. The data logger is set to 3200 Sample/s for a period of 10 s. The experiment is done by holding the data logger and walking on a treadmill at different speeds. Figure 2.8 shows the time domain of the acceleration amplitude for different speed values. The minimum acceleration amplitude at 1.8 mph (2.90 km/h) is 0.41g and the maximum acceleration amplitude at 7 mph is 4.20g, where g is the earth gravity (g = 9.81 m/s^2). The acceleration amplitude increases with increasing the operating speed as the hand vibrates with higher amplitudes at higher operating speeds.

Measured spectra of the data logger output are shown in Fig. 2.9 for 1.8 and 7 mph. As depicted in the figure, 1.8 and 7 mph correspond to 0.857 Hz and 1.476 Hz, respectively. In addition, different harmonic frequencies are shown in Fig. 2.9.

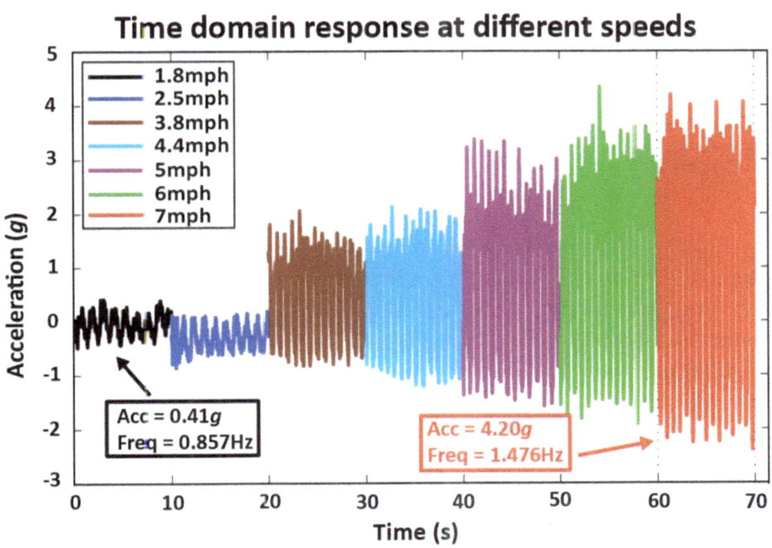

Fig. 2.8 Time domain response at different speeds

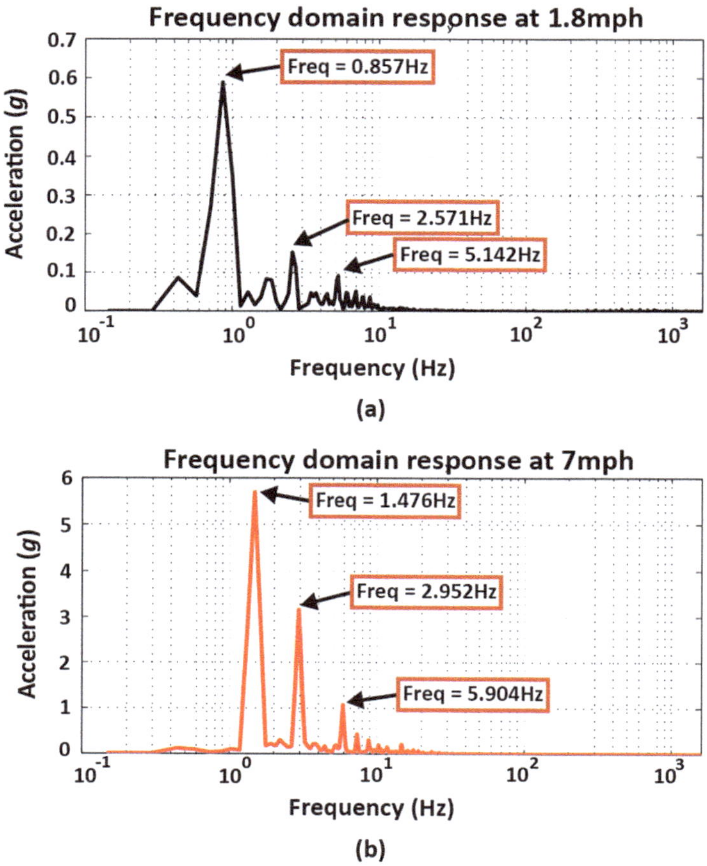

Fig. 2.9 Frequency domain response at (**a**) 1.8 mph and (**b**) 7 mph

Experiment (2): Effect of Tip Mass on V_P at Certain Speed To study the effect of different tip masses on V_P, the experimenter walked on a treadmill at 1.8 mph while monitoring V_P using an oscilloscope. The experiment time is set to 9 s due to a limitation in the oscilloscope configuration at a sampling rate of 25 Samples/s. The measurements are repeated for different tip masses, namely 0, 1.01, 2.10, and 3.14 g. Figure 2.10 shows the time domain output voltage V_P at different tip masses. Inspection of Fig. 2.10 leads to several key observations. The output voltage oscillates around 0 V due to the deflection of the PE beam in both directions while walking. Furthermore, the peak of V_P increases with increasing the tip mass since larger tip mass leads to larger beam deflection, thereby higher output voltage. Further, larger mass value shifts the resonance frequency of the PE harvester to a lower value so that it is closer to the frequency of human vibrations. Finally, V_P reaches its maximum peak around 0.84 V for 3.14 g tip mass.

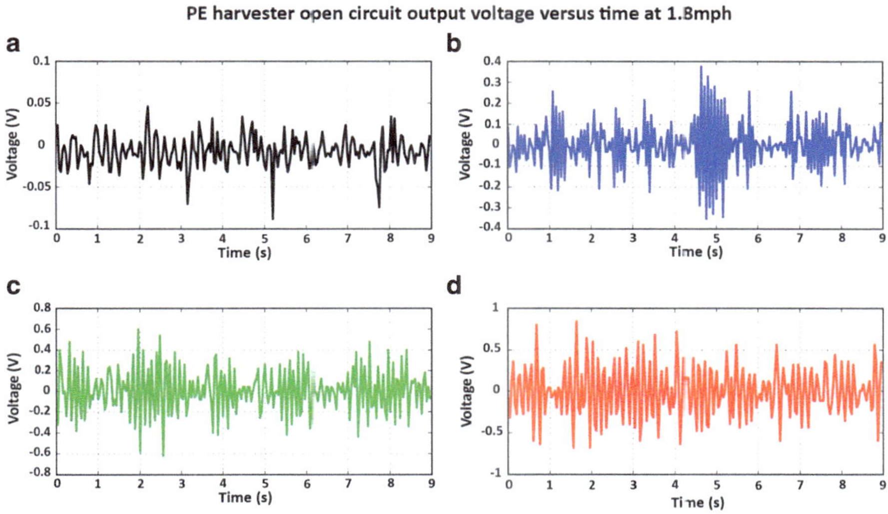

Fig. 2.10 PE harvester open circuit output voltage versus time at 1.8 mph for (**a**) 0 g, (**b**) 1.01 g, (**c**) 2.10 g, and (**d**) 3.14 g

Experiment (3): V_P for 3.14 g Tip Mass at Different Speeds To compare the output voltage V_P at different human speeds, the PE harvester board is fastened on the experimenter's wrist with 3.14 g tip mass. The experimenter ran at multiple speeds on a treadmill while the output voltage is measured using an oscilloscope. Figure 2.11a shows the time domain signal of V_P at seven different speeds, namely 1.8, 2.5, 3.8, 4.4, 5, 6, and 7 mph. As depicted in the figure, V_P increases with increasing the speed because the PE harvester encounters larger deflection at higher speeds, thereby higher output voltage is generated. Furthermore, Fig. 2.11b shows a zoom-in version of V_P at 7 mph. As indicated by the figure, V_P reaches 7.68 V which is one order of magnitude higher than V_P at 1.8 mph.

Experiment (4): Effect of Human Speed on the Rectified Output Voltage To gain an understanding about the harvested energy from the PE harvester, a rectifier circuit is connected to the PE harvester with an output capacitor of 1 μF. As depicted in Fig. 2.12, a simple voltage doubler (VD) circuit rectifies the PE harvester voltage and stores the generated energy in C_{rect}. For simplicity, the piezoelectric equivalent circuit at resonance is used. There are several rectifier circuits reported in the literature such as the Full-Bridge (FB), the VD, the Switch-Only (SO), and the Bias-Flip (BF) rectifiers [49, 50]. In particular, the VD is chosen because it is simple, has a moderate efficiency, and a small area compared to other topologies [51]. In addition, the value of 1 μF is chosen so that the charging time is relatively fast within the time window of the oscilloscope which is 7 s.

Two silicon (Si) diodes (1N4001) are used in the VD circuit with a measured forward voltage drop of 0.54 V. In order to reduce the forward voltage drop,

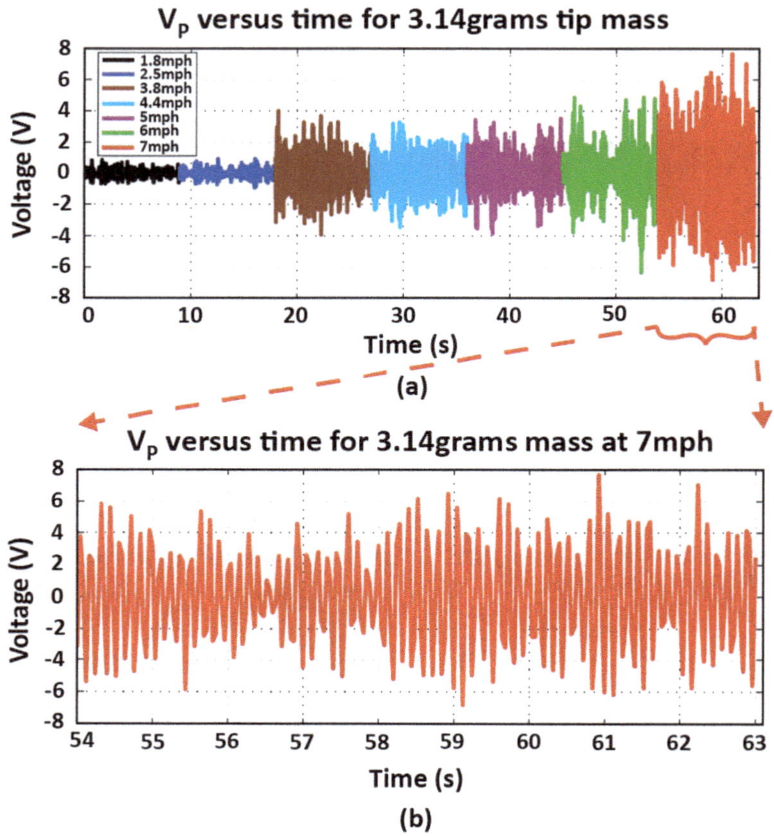

Fig. 2.11 PE harvester open circuit output voltage versus time for 3.14 g tip mass (**a**) at different speeds and (**b**) at 7 mph

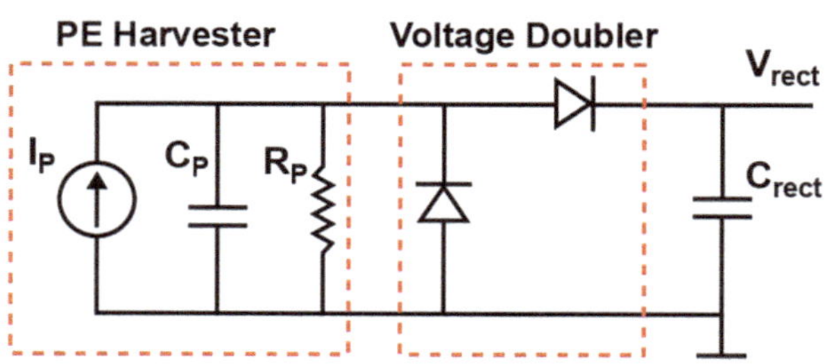

Fig. 2.12 The VD connected to the PE harvester with an output capacitor

germanium (Ge) diodes (OA90) are tested with a measured forward voltage of 0.29 V. However, the Ge diodes have higher reverse current compared to Si diodes, thereby they discharge the output capacitor faster. Therefore, silicon diodes are used in this experiment.

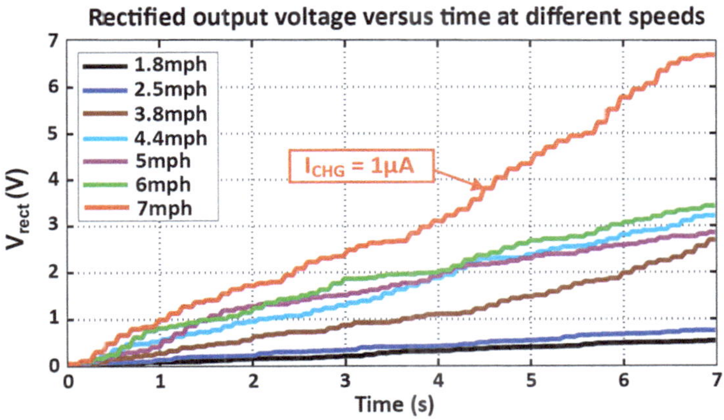

Fig. 2.13 Rectified output voltage versus time at different speeds

Table 2.1 Tabulated harvested energy according to the operated speed

Speed (mph)	Energy (μJ)
1.8	0.14
2.5	0.29
3.8	3.62
4.4	5.19
5	4.09
6	5.99
7	22.40

The output voltage of the VD circuit is measured at different speeds while the PE harvester and the VD are mounted on a human wrist using a breadboard. The output data is measured using the NI ELVIS II board at a sampling rate of 1 kHz [52]. Figure 2.13 shows the time domain signal of the output voltage across C_{rect}. Examination of Fig. 2.13 leads to several key observations regarding the impact of speed on the output voltage. First, we observe the influence of increasing the speed on the output voltage level. In particular, at minimum speed, the output voltage is below 1 V because of the high voltage drop across the diodes. In addition, the charging current at the output capacitor is increasing at higher speed. This corresponds to higher slope in Fig. 2.13 because of larger beam deflection. The larger the beam deflection, the greater the generated charge, and the higher the charging current. Moreover, the charging current at 7 mph is approximately 1 μA, as shown in Fig. 2.13, and the maximum output voltage is around 6.7 V. Finally, Table 2.1 summarizes the stored energy in C_{rect} after 7 s for different speeds.

To extend the measured data in Fig. 2.13 from seconds to hours, the system in Fig. 2.12 is mounted on the human wrist for normal daily activity harvesting. The output capacitor is changed to 470 μF and the V_{rect} is monitored using a microcontroller powered from a separate supply source every 40 s. As depicted

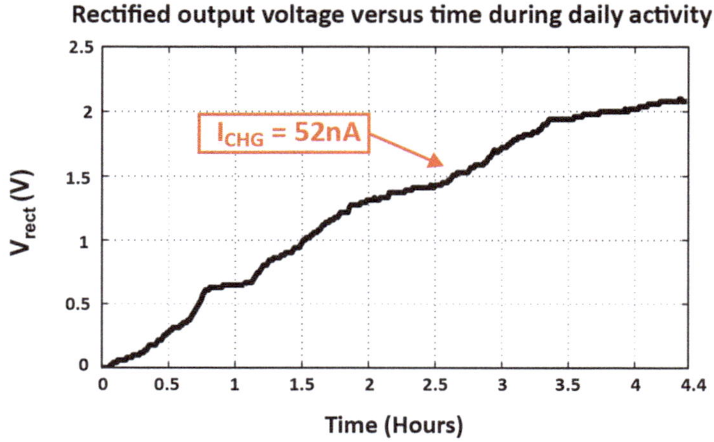

Fig. 2.14 Rectified output voltage versus time in daily activity experiment

in Fig. 2.14, the rectified voltage after 4 h reached more than 2 V. In addition, the amount of energy at the end of the experiment is approximately 1 mJ. Further, the average charge time during the experiment is around 52 nA.

Experiment (5): Output Power at Different Loads Similar to experiment 4, the PE harvester is connected to a VD with an output capacitor C_{rect} of 1 μF. In order to measure the output power ($P_{out(PE)}$), different loads (R_{out}) are connected in parallel with C_{rect}. Moreover, the output voltage (V_{out}) is measured at 7 mph while the PE harvester and the VD are mounted on a human wrist using a breadboard. $P_{out(PE)}$ is calculated from the measured output voltage using Eq. (2.10).

$$P_{out(PE)} = \frac{V_{out}^2}{R_{out}} \tag{2.10}$$

As depicted in Fig. 2.15, the power is maximized at R_{out} of 10 MΩ which is around 3.7 μW. This corresponds to a power density of 7.4 μW/cm^3. Note that the reported power density harvested from human motion could reach 4 μW/cm^3 [53]. However, Mitcheson et al. [53] provided theoretical numbers which vary between different harvesters. Besides, the information provided in [46] is obtained using a shaker that provides consistent vibrations. However, our experiment depends on a human being that moves with inconsistent vibrations. In addition, the loss in the interface circuit is different between our experiment and the one in [46].

To extract the maximum power from the PE harvester, a conjugate mismatch circuit should be implemented. This circuit consists of a large inductor to cancel the effect of the parasitic capacitor, and a resistor that matches the value of R_P [49]. In addition, MPPT circuit can be implemented after the rectifier circuit as explained in [54]. In our setup, the large inductor is impractical for wearable applications.

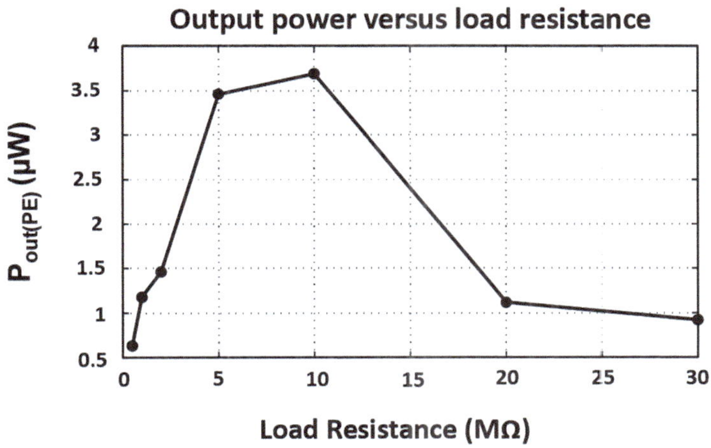

Fig. 2.15 Output power versus output load at 7 mph

Fig. 2.16 Typical block diagram for RF harvester

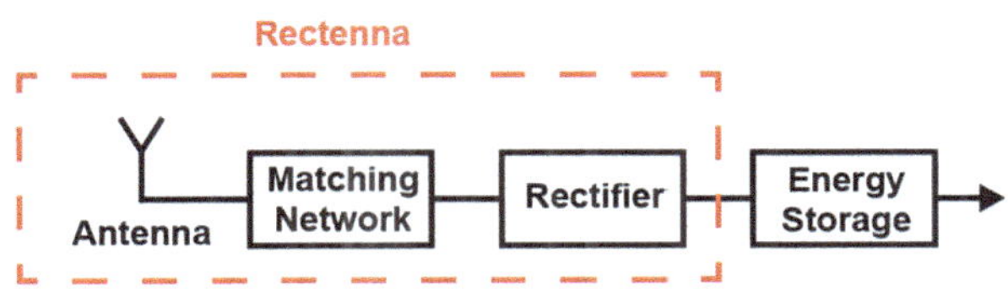

Further, MPPT circuit is not used in order to simplify the experiment. However, future improvements should be implemented to further maximize the efficiency of the system.

2.1.3 RF Harvesting

Radio frequency (RF) harvesting is achieved by scavenging some of the energy in certain RF bands and converting it into electricity. The most common RF bands for energy harvesting are the 900 MHz, which is the GSM band, and the 2.4 GHz, which is the industrial, scientific, and medical (ISM) band. Figure 2.16 shows a typical block diagram on how to convert the RF energy into an electrical energy. The antenna is the interface with the RF signal and it needs to be designed to have high gain at the band of interest. The output from the antenna is an electrical AC signal at the radio frequency.

The power available at the antenna (P_{AV}) can be calculated using Eq. (2.11), where S is the power density at the receiver side and A_e is the effective aperture of the antenna. S and A_e can be also calculated using Eqs. (2.12) and (2.13), respectively, where P_{EIRP} is the effective isotropic radiated power, G_r is the antenna gain of the receiver, d is the distance between the transmitter and receiver, and λ is the wavelength of the radio signal. S is a function of the distance, as the distance gets smaller, the received power density becomes higher. Further, the available power at the antenna is a function of the frequency, where the higher the frequency the lower the power. However, there is a trade-off between the operating frequency and the antenna size where the higher the frequency the smaller the antenna can be.

$$P_{\mathrm{AV}} = S \cdot A_e \tag{2.11}$$

$$S = P_{\mathrm{EIRP}} \frac{1}{4\pi d^2} \tag{2.12}$$

$$A_e = \frac{G_r \lambda^2}{4\pi} \tag{2.13}$$

The antenna design is an important factor to achieve good power and efficiency. Figure 2.17 shows the electrical model of an antenna. As depicted in the figure, the antenna can be described by an AC voltage source in series with an impedance of the antenna. The impedance consists of a resistor and a reactance. The voltage generated across the antenna can be calculated using Eq. (2.14), where R_s is the series resistance of the antenna. V_s is preferred to be high so that the efficiency of the circuit afterward can be also good. In addition, V_s can be increased by using multiple antennas and connect them in series. In contrast, if the antennas are connected in parallel, the current will be increased.

$$V_s = \sqrt{2R_s P_{\mathrm{AV}}} \tag{2.14}$$

In addition to the antenna, Fig. 2.16 shows a matching network, rectifier circuit, and energy storage. The matching network matches the impedance of the antenna to achieve maximum power transfer. The rectifier circuit transforms the AC into DC signal that can be then stored in an energy storage. The antenna, matching network, and the rectifier together are called rectenna which is a rectifying antenna (Fig. 2.16).

The importance of the impedance matching block comes from maximizing the power transfer from the antenna to the rectifier which in result increases the conversion efficiency. This matching circuit is designed to match the impedance of the antenna 50 Ω with the input impedance of rectifier circuit. There are three ways the matching circuit can be implemented: L matching network, π matching network,

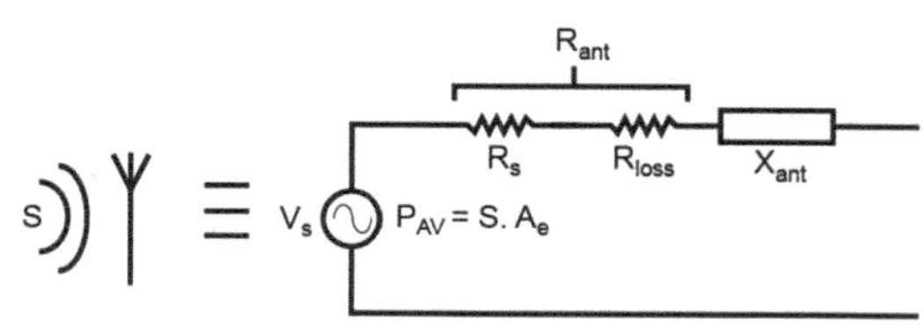

Fig. 2.17 A general electrical model of an antenna

Table 2.2 Matching technique alternatives

Topology	L-network	π-Network	T-network
Simplicity	Simplest (two components)	More complex (three components)	More complex (three components)
Power loss	Least power loss	More power loss	More power loss
Quality factor	Note controllable	Controllable	Controllable

Fig. 2.18 The RF harvester, P2110CSR-EVB evaluation board

and *T* matching network. The *L* matching technique is the most preferred technique in RF energy harvester since it has only two lumped elements, a capacitor, and an inductor which in result minimizes the losses in the circuit [55]. Table 2.2 shows a brief comparison between the different matching techniques.

2.1.3.1 Characterization of RF Energy Harvesting at 2.4 GHz

This section addresses the characterization of an RF energy harvesting system including four different types of antennas dedicated for 2.4 GHz by using P21XXCSR-EVB RF harvesting kit from Powercast. This kit can harvest from six different bands including the targeted 2.4 GHz band. The goal is to study the effect of different types of antenna on the output power, efficiency, and output voltage. Also, the work includes changing the distance between the transmitting source and the receiver to study the effect on the received power.

RF Harvesting Kit The kit P2110CSR-EVB shown in Fig. 2.18 has control capabilities of many parameters. There are six frequency bands to harvest from: GSM-850 uplink, Europe RFID and GSM-850 downlink, ISM USA and GSM-900 uplink, GSM-1800, GSM-1900 uplink, and Wi-Fi 2.4 GHz. The user can place a storage capacitor or choose one of the two storage capacitors in the kit: 2200 mF electrolytic capacitor and 50 mF super capacitors. There is a sliding switch that allows varying the regulated output voltage between 3.4, 4.1, and 4.2 V. The 50 mF capacitor is used for storage, and two regulated voltages are selected 3.4 and 4.1 V.

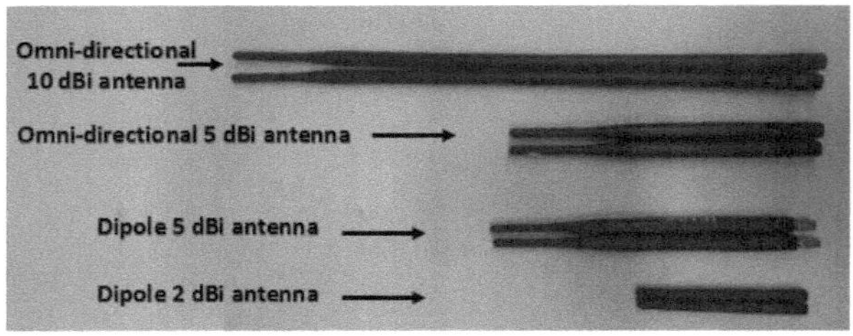

Fig. 2.19 Antennas used in RF harvester characterization

Antennas As depicted in Fig. 2.19, four different antennas are used in the experiments: 2 dBi dipole antenna, 5 dBi dipole antenna, 5 dBi omni-directional antenna, and 10 dBi omni-directional antenna. Different antennas are experimented as the gain and the type of antenna play an important role in extracting the maximum RF power from the surrounding.

Low RF-Input Power Experiment This experiment is conducted to find the maximum distance at which the RF harvesting kit can regulate an output voltage of 3.4 V. Two antennas of the same type are used each time. One is the transmitting antenna that is connected to the RF signal generator. The second is the receiving antenna and it is connected to the 2.4 GHz-specified SMA connector of the P21XXCSR-EVB evaluation board. The board is distanced gradually from the transmitting antenna while the output voltage is measured and observed simultaneously. Figure 2.20 shows the graph of the maximum distance at which each antenna can harvest when varying the input power of the RF signal generator at a fixed frequency of 2.4 GHz. Based on the results obtained in Fig. 2.20, the maximum distance at which the board can regulate an output voltage of 3.4 V is 116 cm away from the transmitting antenna using a 10 dBi omni-directional antenna and an input power of 21 dBm at 2.4 GHz. In this experiment, the gain of the antenna is the determinant parameter as the antenna with the highest gain outperformed the other antennas at longer distances regardless of their type.

Maximum Output Power Experiment In this experiment, a varying load (potentiometer) is connected across the output terminals of the evaluation board. The distance from the transmitting antenna is fixed at 10 cm and the load is varied until the minimum resistance before the output is overloaded. This process is repeated for each type of the antennas obtained and for different RF input power values. Table 2.3 demonstrates the maximum output power at which the board can regulate an output voltage of 3.4 V using different types of antennas and different input power at 2.4 GHz.

Figure 2.21 shows the maximum output power delivered to the load when the input power is varied and using different types of antennas 10 cm away from the transmitting antenna at 2.4 GHz. As demonstrated in Fig. 2.21, the maximum output power that can be delivered to the load is 4 mW using a 5 dBi dipole antenna with an input power of 21 dBm at 2.4 GHz.

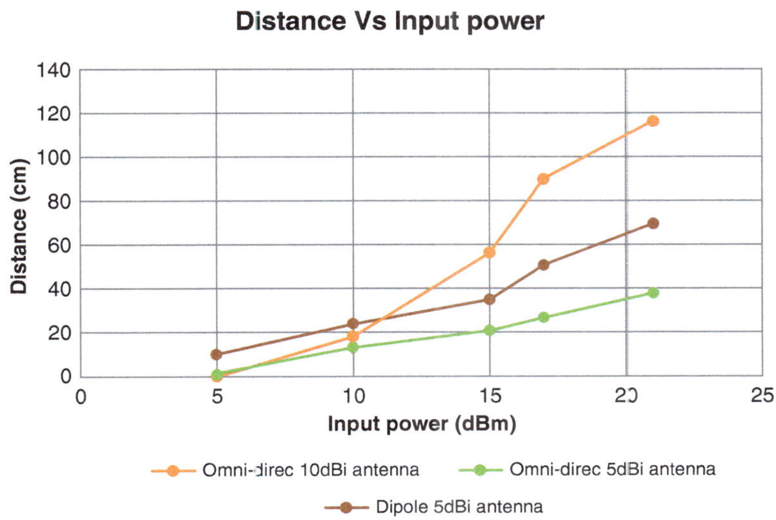

Fig. 2.20 Measurement of the maximum distance at which different antennas can harvest when varying the input power at 2.4 GHz

Table 2.3 Maximum output power at which the evaluation board can regulate 3.4 V at different input powers using different antennas when harvesting at 10 cm away from the source

	Input power (dBm)					
	3	6	9	12	15	21
Antenna	Output power (μW)					
Omni-directional 10 dBi	0	0	13	116.4	312.4	1445
Omni-directional 5 dBi	0	0	23	81	323.8	1298.8
Dipole 5 dBi	1.2	50.3	165.1	444.6	963	4000
Dipole 5 dBi	0	0	0	48.8	112.7	722.5

An attempt to extract the maximum output power of the harvesting board at a regulated output voltage of 4.1 V is also conducted. The receiving antenna is placed at 10 cm away from the transmitting antenna and the experiment is conducted for the most outperforming antennas obtained for different levels of input power at 2.4 GHz. Figure 2.22 shows the maximum output power delivered to the load at 4.1 V regulated voltage. As shown in Fig. 2.22, the maximum output power that can be delivered to the load at a regulated voltage of 4.1 V is 3.5 mW using a 5-dBi dipole antenna. The results show that despite the dipole antenna has a lower gain than the omni-directional antenna, it outperformed the omni-directional antenna at shorter distances. This can be regarded to the different radiation pattern of different types of antennas as the effect of the antenna radiation pattern can be observed clearly at shorter distances.

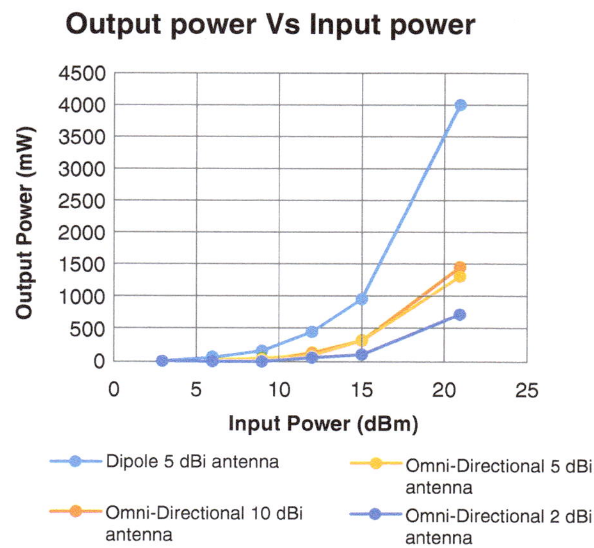

Fig. 2.21 Measurement of maximum output power delivered to the load at different input power using different types of antennas

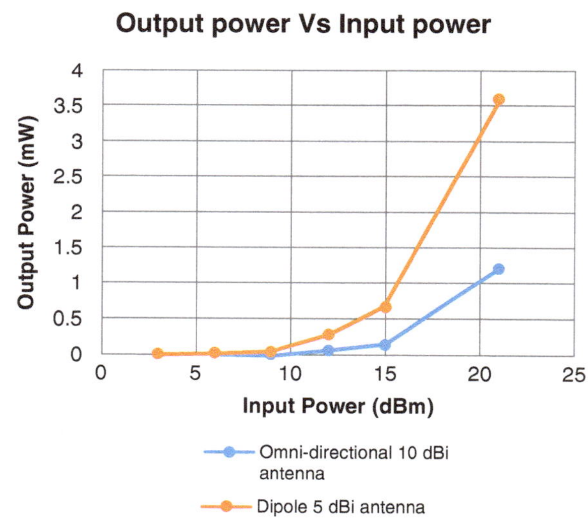

Fig. 2.22 Measurements of the maximum output power delivered to the load at a regulated output voltage of 4.1 V

Efficiency of the System and Power Losses The total power efficiency, η, is the ratio of the DC power at the output of the harvester to the power of the RF input signal from the RF signal generator. The efficiency is calculated using the measured values of the output power obtained in the second experiment. This end-to-end efficiency of the RF harvesting system is reported in Fig. 2.23. As depicted in Fig. 2.23, the total efficiency of the system increases when the input power is increased. The maximum total efficiency of the RF harvesting system obtained is 3.18% at an input power

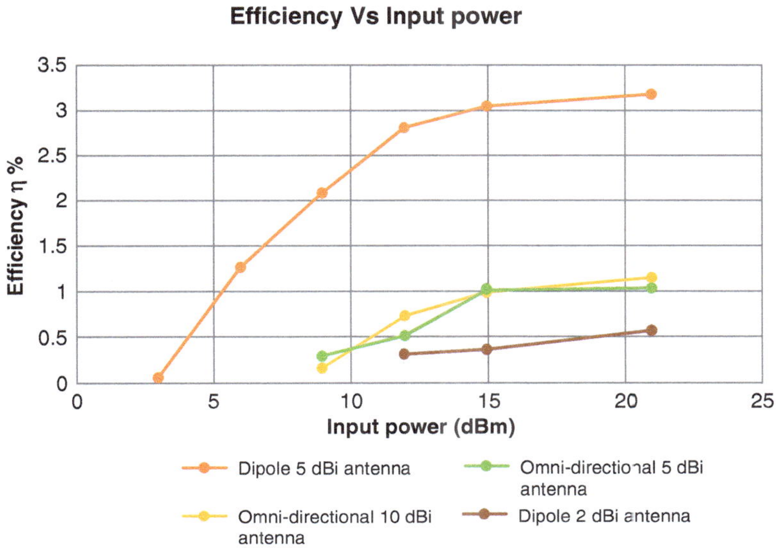

Fig. 2.23 End-to-end efficiency of the RF energy harvesting system (Total efficiency)

of 21 dBm. The value obtained indicates a high effect of losses on the efficiency of the RF harvesting system. The losses of the system can be classified into three categories: losses in the harvester kit, losses in the air between transmitting and receiving antenna, and losses within the transmitting antenna as part of the energy transmitted does not reach the receiving antenna. The total efficiency of the system can be calculated using Eq. (2.15), where η_1 is the efficiency of the harvester kit and η_2 is the efficiency of the other losses of the system (considering losses in the air and the antennas).

$$\eta_{\text{total}} = \eta_1 \times \eta_2 \qquad (2.15)$$

Based on the datasheet of the evaluation board, the efficiency of the harvester kit at 15 dBm input power is 50%. Given $\eta_{\text{total}} = 3\%$ at 15 dBm from Fig. 2.23 and η_1 from the datasheet, η_2 can be calculated using Eq. (2.15) and it is equal to 6%. The obtained value of η_2 indicates that the efficiency of the RF harvesting system is mainly affected by the losses in the air and the antennas. The power transmitted and received by the antennas is affected by several parameters as indicated in Eq. (2.16) by Friis' transmission equation. Where P_t is the transmitted power, P_r is the received power, G_t is the gain of the transmitting antenna, G_r is the gain of the receiving antenna, λ is the wavelength, and d is the distance between the transmitter and the receiver.

$$P_r = \frac{P_t G_t G_r \lambda^2}{(4\pi d)^2} \tag{2.16}$$

Equation (2.16) shows the effect of the distance between the transmitting and receiving antennas and the gain of the antennas on the received and transmitted power. The power received decreases with increasing distance and increases with increasing gain. Hence, it can be concluded that most of the power losses in the RF energy harvesting system experimented are due to losses in the air and the antennas. Therefore, it is recommended to use highly efficient power transmitting and receiving antennas to achieve better RF energy scavenging.

Further improvement of the RF harvesting system can be achieved by using antenna arrays. The use of antenna arrays has been suggested as a way to increase the efficiency of RF energy harvesting systems [56–58]. An array of narrow band antennas can increase the amount of RF energy harvested by each antenna. Consequently, this increases the amount of power harvested at a specific band by combining the energy harvested by each narrow band antenna [58], hence, increasing the power extraction efficiency of the system. In addition, antenna arrays could be effective in increasing the efficiency of harvesting at low input power [56].

2.1.4 Solar Harvesting

Solar energy harvesting is based on photoelectric effect which was discovered by Albert Einstein and received the Nobel Prize for such discovery in 1921. The photoelectric effect is the generation of an electron–hole pair in a semiconductor material due to the absorption of a photon. The electron–hole pair is generated if the photon energy is larger than the bandgap energy of the semiconductor material. Equation (2.17) shows the energy of the single photon, where h is Planck's constant, c is the speed of light, and λ is the wavelength of the incident light. Figure 2.24 shows the operating principle of a solar cell. As depicted in the figure, when the light hits a *p–n* semiconductor, then an *e–h* pair is generated and collected at the load. An anti-reflection material can be added to the solar cell in order to increase the absorption of the incident photons.

$$E = \frac{hc}{\lambda} \tag{2.17}$$

Figure 2.25 shows the electrical model of a solar cell. It consists of a current source, a diode, parallel and series resistors. The diode supplies the output voltage while the current source provides a constant current to the load over the range of output voltage. The parallel resistance models the leakage current loss in the diode. Further, the series resistance models the resistance of the bulk and the metallic contacts.

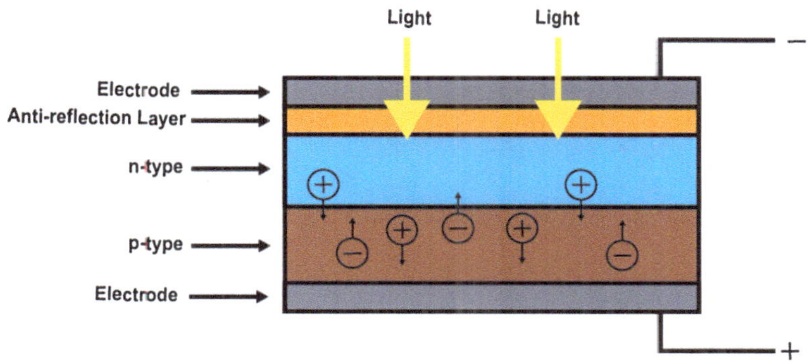

Fig. 2.24 Operating principle of solar cell

Fig. 2.25 A common electrical model of a solar cell

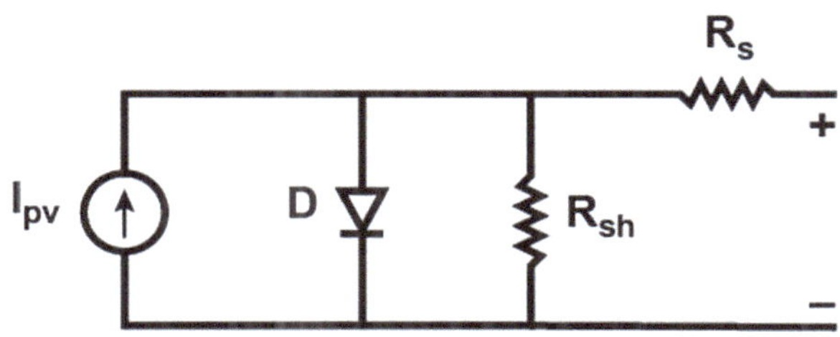

Fig. 2.26 *I–V* characteristic of a solar cell

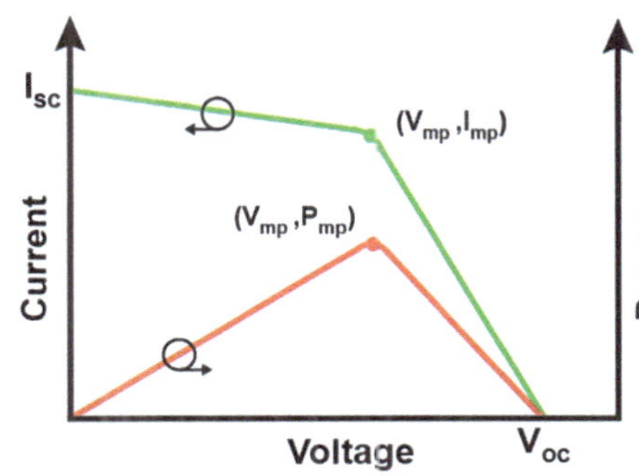

The *I–V* characteristic of the solar cell is shown in Fig. 2.26. As depicted in the figure, there are three important parameters that define the performance of a solar cell: open circuit voltage (V_{oc}), short circuit current (I_{sh}), and maximum power point (P_{mp}). V_{oc} is the maximum output voltage from the solar cell at no load. I_{sh} is the maximum current from the solar cell when the output is shorted. P_{mp} is the maximum power that can be extracted from the solar cell at the operating output voltage and current (V_{pm},I_{pm}).

Fill factor (FF) of a solar cell is the ratio between the maximum output power to the ideal maximum output power as shown in Eq. (2.18). The larger the value of

FF the better the solar cell. Finally, the efficiency of the solar cell is defined as the maximum output power divided by the input power as shown in Eq. (2.19)

$$\mathrm{FF} = \frac{V_{\mathrm{mp}} I_{\mathrm{mp}}}{V_{\mathrm{oc}} I_{\mathrm{sh}}} \tag{2.18}$$

$$\eta = \frac{V_{\mathrm{mp}} I_{\mathrm{mp}}}{P_{\mathrm{in}}} \tag{2.19}$$

2.2 Power Conversion Circuits

Power conversion circuits are the interface between the energy harvester and the load. These circuits convert the voltage generated from the harvesters into a level that is suitable for the load. Further, they generate multiple, regulated output voltages that power different modules in the load. For example, in typical SoC, the processor runs at low supply voltage down to 0.5 V. However, the SRAM needs to have higher than 0.8 V to operate. In addition, wireless circuits usually work at higher voltage supply, $\sim$1 V. Hence, the power conversion circuit has to support enough power while generated multiple output voltages.

Power conversion circuits typically include AC–DC converters and DC–DC converters. Figure 2.27 shows a general block diagram of power conversion circuits in energy harvesting systems [59]. It includes an AC–DC converter for vibrational and RF energy harvesting since the nature of the signal is AC. Then, a DC–DC converter is used to convert the voltage into a different level. Further, an intermediate storage element such as a battery or a super capacitor is needed. Finally, a power management block is necessary so that it drives different loads with suitable voltage levels. In the following subsections, different types of DC–DC converters will be introduced that are commonly used in energy harvesting applications.

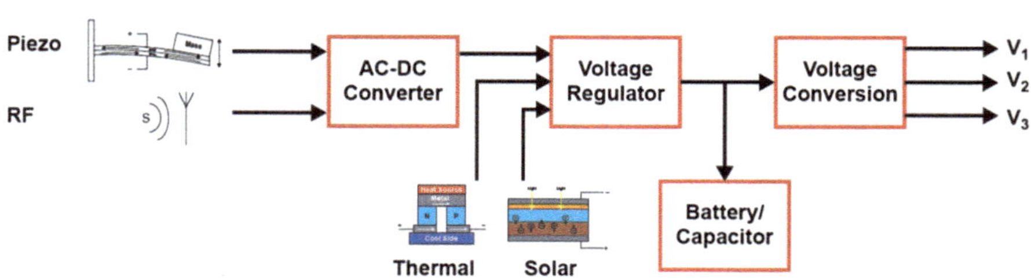

Fig. 2.27 A general energy harvesting system for multiple energy sources

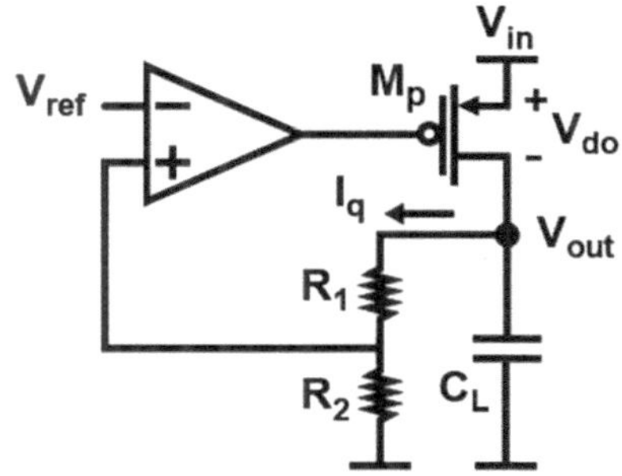

Fig. 2.28 Illustration of a linear regulator circuit

2.2.1 *Linear Regulators*

Linear regulators (LDs) are widely utilized in supplying circuits of multiple voltage domains in many power management chips [60]. LDs offer the advantages of a fully on-chip integrated converter with very low area consumption and good achievable ripple and transient characteristics. Figure 2.28 shows an illustration of an LD basic circuit. As depicted in the figure, a control feedback is required to adjust the resistance of the pass transistor so that the output voltage is regulated. The output voltage depends on the voltage division across R_1 and R_2 as shown in Eq. (2.20).

$$V_L = V_{\text{ref}}\left(1 + \frac{R_1}{R_2}\right) \tag{2.20}$$

Equation (2.20) leads into several key information regarding the LD characteristics. First, the output voltage is scaled linearly with the input voltage which also implies the same for the efficiency. Further, the LD is not able to boost the voltage, it works only as a buck converter. As such, the LD is not suitable for applications where the output needs to be larger than the input voltage. However, it can be used as a second stage for regulating the output voltage. The voltage drop on the pass transistor M_p is defined as shown in Eq. (2.21). Further, the efficiency is defined as shown in Eq. (2.22), where I_q is the static current going through the voltage divider and the leakage into the opamp.

$$V_{\text{do}} = V_{\text{in}} - V_{\text{out}} \tag{2.21}$$

$$\eta = \frac{P_{\text{out}}}{P_{\text{in}}} = \frac{V_{\text{out}} I_{\text{out}}}{V_{\text{in}} I_{\text{in}}} = \frac{(V_{\text{in}} - V_{\text{do}}) I_{\text{out}}}{V_{\text{in}}(I_{\text{out}} + I_q)} \tag{2.22}$$

2.2.2 *Switched-Capacitor Circuit*

Switched-capacitor (SC) converters are used where high voltages, or opposite polarity, than the input is needed. It uses capacitors and switches to transfer the charge from one capacitor to another as shown in Fig. 2.29. In addition, SC capacitor

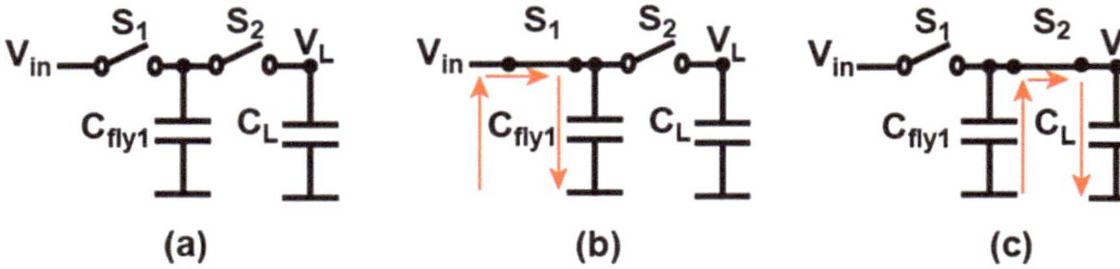

Fig. 2.29 (**a**) Illustration of SC circuit during (**b**) charging phase and (**c**) discharging phase

can be configured to generate multiple output voltage by changing the ratio between capacitors [61]. SC provides the advantage of an integrated solution without the need of any off-chip components. In addition, it has higher efficiency than the LD. However, SC circuits suffer from the following issues. First, it is difficult to maintain the efficiency across wide load voltages. Further, the output current capability is limited. Second, to scale up the voltage, many switched capacitors need to be connected in series to boost the voltage [62]. This results into degrading the efficiency and complex design in controlling many SC circuits. Moreover, there is an inherent loss in the SC circuit when the charge is transferred from one capacitor to another which prevents the efficiency to be 100% even if all components are ideal [63].

There are four main losses contributors in SC circuits: conduction losses (P_{cond}), switching losses (P_{sw}), flying capacitor losses ($P_{c_{\text{fly}}}$), and bottom plate parasitic losses (P_{bp}). P_{cond} losses are due to the on resistance of the switches as shown in Eq. (2.23), where R_{ds} is the drain-source resistance of the transistors, and I_{ds} is the drain-source current in the transistor. P_{sw} losses are due to charging and discharging the capacitance at the transistor gate as shown in Eq. (2.24), where C_{ox} is the oxide capacitance, V_{in} is the input voltage, W is the transistor width, and f_{sw} is the switching frequency. ($P_{c_{\text{fly}}}$) loss is shown in Eq. (2.25) which is due to charging and discharging the flying capacitors, where I_c is the current passing through C_{fly}. Finally, (P_{bp}) loss is due to the existence of parasitic capacitance which results into losses during transition as shown in Eq. (2.26)

$$P_{\text{cond}} = I_{\text{ds}}^2 R_{\text{ds}} \tag{2.23}$$

$$P_{\text{sw}} = C_{\text{ox}} V_{\text{in}}^2 W f_{\text{sw}} \tag{2.24}$$

$$P_{c_{\text{fly}}} = \frac{I_c^2}{C_{\text{fly}} f_{\text{sw}}} \tag{2.25}$$

$$P_{\text{bp}} = C_{\text{bp}} V_{\text{in}}^2 f_{\text{sw}} \tag{2.26}$$

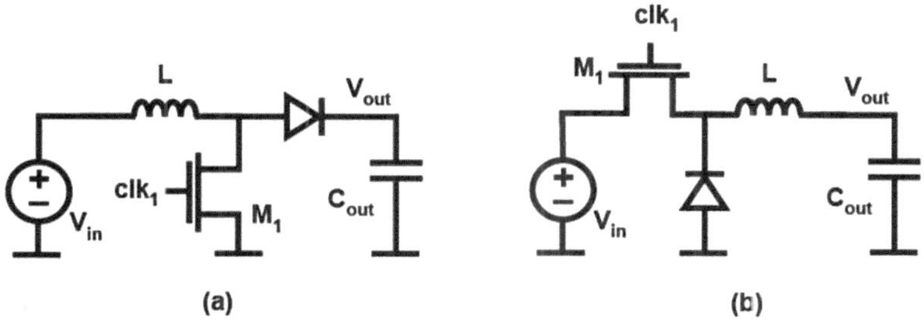

Fig. 2.30 A basic circuit of inductor-based (**a**) boost and (**b**) buck converter

Table 2.4 A comparison between different types of converters

Comparison	Linear regulator	Switched capacitor	Inductor converter
Fully integrated	✓	✓	X
High efficiency	X	✓	✓✓
High step-up voltage gain	X	X	✓

2.2.3 *Switching Converters*

An inductor-based DC–DC converter is one of the most efficient converters. It uses an inductor to store the energy and then release it to the load. Theoretically, the inductor-based converter can achieve 100% efficiency regardless of the output load. The switching mechanism in the converter can be controlled by a digital circuit which could consume low power [64]. The inductor converter can have a very high voltage gain which makes it more suitable for energy harvesting applications. There are two types of inductor converters: boost and buck converter as shown in Fig. 2.30. Boost converter converts the low input voltage to a higher one while the buck converter does the opposite. Both boost and buck have a similar design as shown in Fig. 2.30b. The main drawback is the usage of the off-chip components (i.e., inductor), which increases the size of the system. Fortunately, advance fabrication process allows small form factor for off-chip components which could be adequate for portable applications.

Table 2.4 shows a comparison between the different types of converters explained in the last sections. As depicted in the figure, inductor-based DC–DC converter achieves high efficiency with high gain which makes it suitable for TEG harvesting applications.

Chapter 3
Interface Circuits for Thermoelectric Generator

3.1 Inductor-Based DC–DC Converters

An inductor-based DC–DC converter is one of the most efficient converters. It is widely used in energy harvesting systems especially the TEG harvesting applications. There are two types of inductor-based DC–DC converters, namely asynchronous and synchronous converters. The following subsections explain each in detail.

3.1.1 An Asynchronous Inductor-Based DC–DC Converter

Figure 3.1a shows the circuit diagram of an asynchronous inductor-based boost converter. It consists of an inductor, a switch M_1, a diode D_1, input and output capacitors. The operation of the circuit is as follows. When clk_1 is high, then M_1 is turned on and the inductor is charged from the input voltage as shown in Fig. 3.1b (assuming ideal inductor). The charging current is shown in Eq (3.1), where t_{on} is the ON period of clk_1.

$$I_{\text{on}} = t_{\text{on}} \frac{V_{\text{in}}}{L} \tag{3.1}$$

In contrast, when clk_1 is low, the inductor flips its voltage polarity and the discharging current is shown in Eq. (3.2).

$$I_{\text{on}} = t_{\text{off}} \frac{V_{\text{in}} - V_{\text{out}}}{L} \tag{3.2}$$

M. Alhawari et al., *Energy Harvesting for Self-Powered Wearable Devices*, Analog Circuits and Signal Processing, DOI 10.1007/978-3-319-62578-2_3

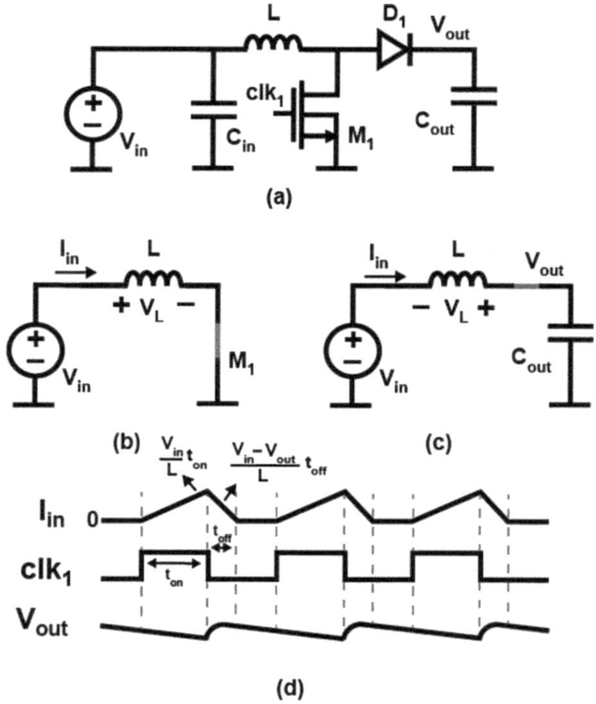

Fig. 3.1 (**a**) An asynchronous inductor-based boost converter. (**b**) Charging phase. (**c**) Discharging phase. (**d**) Waveforms show the converter operation

Based on the volt-second balance principle of the inductor, $I_{on} = I_{off}$ at steady state. Equating Eqs. (3.1) and (3.2), then the output voltage of the boost converter can be calculated under steady state conditions as shown in Eq. (3.3) [65]. where D is the duty cycle of clk_1.

$$V_{out} = \frac{V_{in}}{(1 - D)} \tag{3.3}$$

Inspecting of Eq. (3.3) reveals some key information regarding the asynchronous design considerations. First, the implementation of the asynchronous inductor-based boost converter is fairly easy where only one clock is used and its duty cycle affects the desired voltage gain. However, for high voltage gain, the duty cycle needs to be large. For example, for a gain of 50, which is typical in TEG application, the duty cycle needs to be 98%. This puts a design constraint on the converter where such duty cycle might not be feasible. Further, the usage of the diode degrades the efficiency due to the diode voltage drop and its internal resistance. As such, the trade-off here is between the efficiency of the converter and the complexity of the control circuit given the other specifications such as the voltage gain.

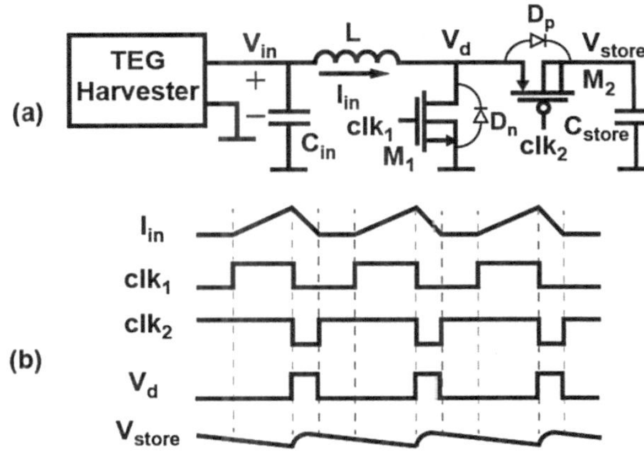

Fig. 3.2 (**a**) Circuit diagram of a synchronous inductor-based boost converter. (**b**) Voltage and current waveforms of the converter during operation

3.1.2 A Synchronous Inductor-Based DC–DC Converter

The main difference between the asynchronous and the synchronous converter is in the high side switch. The synchronous converter uses a PMOS switch instead of a diode to reduce the voltage drop and enhance the efficiency of the converter. Figure 3.2a shows the circuit implementation of the synchronous inductor-based boost converter. As depicted in Fig. 3.2b, the inductor is charged when clk_1 is high through the switch M_1. Then, the inductor releases its stored energy to the output capacitor when clk_2 is low. Note that the falling edges of clk_1 and clk_2 are synchronized so that clk_2 is low as soon as clk_1 becomes low. The deadtime between the clocks should be as small as possible in order to reduce the spike on the V_d node and to prevent turning on the parasitic diode of the PMOS switch.

The gain of the synchronous converter can be derived the same way as the asynchronous one as shown in Eq. (3.4). Where $T_{\text{nmos,on}}$ and $T_{\text{pmos,on}}$ are the on-time for the NMOS and PMOS switches, respectively.

$$\frac{V_{\text{out}}}{V_{\text{in}}} = 1 + \frac{T_{\text{nmos,on}}}{T_{\text{pmos,on}}} \tag{3.4}$$

In low power applications, it is more efficient to run the converter in discontinuous current mode (DCM). As such, $T_{\text{pmos,on}}$ is the time till the inductor current reaches zero. The input capacitance C_{in} needs to be large enough so that it holds the TEG voltage at the operating voltage. This is useful when performing maximum power point transfer. For the output capacitor, it needs to be selected to filter the switching noise generated from the converter operation.

3.2 Design of the Synchronous Inductor-Based Boost Converter

One approach to design the synchronous inductor-based DC–DC converter is by first choosing the inductance value. A higher inductance gives lower losses for the same source impedance [66]. As such, there is a trade-off between the form factor of the inductor and the efficiency. In addition, the inductance value affects the impedance matching with the harvester.

The voltage gain of the converter is an important factor in TEG harvesting since the input voltage range is small (20–50 mV). Different losses in the inductor-based DC–DC converter (inductor quality factor, conduction, and switching losses) limit the gain of the converter where the highest gain happens at light load.

Typically, a low duty cycle for the clock of the converter is preferred in low power applications. However, a duty cycle of 0.5 can be chosen for the following reasons. First, the implementation of a 0.5 duty cycle clock generator is fairly easy where a ring oscillator can be used. In contrast, lower duty cycle circuit requires extra circuits which contribute to the cost of the overall power consumption [67]. Second, for a fixed inductance and TEG series resistance (R_{TEG}), decreasing the charging phase (lower duty cycle) of the inductor, $T_{\text{nmos,on}}$, will require lowering the discharging phase $T_{\text{pmos,on}}$ assuming a fixed ratio in Eq. (3.4) for a fixed voltage gain. As such, the zero current switching (ZCS) circuit needs to be faster for lower $T_{\text{pmos,on}}$ so that it turns off the PMOS switch faster which makes the design of the ZCS circuit elusive. More discussion will be introduced in Chap. 4.

After choosing the duty cycle of the clock, the clock frequency of the inductor-based DC–DC converter can be calculated from Eq. (3.5) such that the maximum power transfer is achieved, where D is the duty cycle of the clock. For example, the clock frequency at D equals to 0.5, R_{TEG} of 7 Ω, and L of 47 μH is ∼18 kHz.

$$f_{clk_1} = \frac{R_{\text{TEG}} x D^2}{2L} \qquad (3.5)$$

3.2.1 *Losses in the Inductor-Based Boost Converter*

Theoretically, the inductor-based converters can achieve 100% efficiency if there are no losses. However, losses contribute in degrading the efficiency and the gain of the converter. As such, a good understanding of different losses helps in overcoming some of the typical mistakes in designing the inductor-based converters. The major losses in the inductor boost converter are the conduction losses from the NMOS, PMOS switches, and the inductor series resistance. In addition, the gate capacitances of the NMOS and PMOS switch with the parasitic capacitances from the printed circuit board contribute in the overall switching losses. Finally, the leakage current also degrades the efficiency, especially in low power applications. The total loss in the boost converter can be approximated using Eq. (3.6) [15].

$$\begin{aligned}\text{Total-loss} = &\left[C_{\text{GS,NMOS}} + C_{\text{GD,NMOS}} \cdot \left(1 + \frac{V_{\text{in}}}{V_{\text{out}}}\right)\right] \cdot W_{\text{NMOS}} \cdot (V_{\text{out}})^2 \cdot f_{clk} \\ &+ \frac{1}{\eta} \cdot [C_{\text{GS,PMOS}} + C_{\text{GD,PMOS}}] \cdot W_{\text{PMOS}} \cdot (V_{\text{drive}})^2 \cdot f_{clk} \\ &+ V_{\text{out}} \cdot I_{\text{leak,NMOS}} \cdot W_{\text{NMOS}} + \frac{1}{\eta} \cdot V_{\text{drive}} \cdot I_{\text{leak,PMOS}} \cdot W_{\text{PMOS}} \\ &+ \frac{I_p^2}{3} \cdot T_{\text{NMOS,on}} \cdot f_{clk} \cdot \frac{R_{\text{NMOS}}}{W_{\text{NMOS}}} + \frac{I_p^2}{3} \cdot T_{\text{PMOS,on}} \cdot f_{clk} \cdot \frac{R_{\text{PMOS}}}{W_{\text{PMOS}}} \\ &+ V_{\text{in}} \cdot I_{\text{leak,NMOS}} \cdot W_{\text{NMOS}} + (V_{\text{out}} - V_{\text{in}}) \cdot I_{\text{leak,PMOS}} \cdot W_{\text{PMOS}} \\ &+ \frac{1}{2} \cdot C_{\text{eff}} \cdot V_{\text{out}}^2 \cdot f_{clk} + \frac{I_p^2}{3} \cdot (T_{\text{nmos,on}} + T_{\text{pmos,on}}) \cdot f_{clk} \cdot R_{L,\text{ESR}} \qquad (3.6)\end{aligned}$$

where $C_{\text{GS,NMOS}}$, $C_{\text{GD,NMOS}}$, $C_{\text{GS,PMOS}}$, and $C_{\text{GD,PMOS}}$ are the gate to source and gate to drain capacitances per unit width for the NMOS and PMOS switches, I_p is the peak inductor current, R_{NMOS} and R_{PMOS} are the resistances of the switches, C_{eff} is the total capacitance at the common node between the NMOS and the PMOS. Note that the Miller effect appears for the NMOS switch only [15, 65].

Equation (3.6) reveals key information regarding the parameters that affect the converter efficiency. There are many parameters where optimization needs to be done between them. SPICE simulation helps in optimizing the converter with regard to the efficiency. For example, for choosing the width of the switches, a sweep can be done versus the efficiency. Then the width of the switches can be chosen when the efficiency is maximized.

For example, Fig. 3.3 shows a SPICE simulation in 65 nm CMOS for a synchronous inductor-based converter assuming both switches have the same width. As depicted in the figure, the efficiency is calculated at different width values as well as load resistances. Hence, the optimized width for the switches is 10 mm since the efficiency saturates for widths larger than 10 mm. The maximum efficiency is achieved at 15 kΩ load resistance.

3.2.2 Control Circuits for Inductor-Based Converter

There are many digital control circuits for the inductor-based boost converter that are necessary for proper operation. In this book, the focus will be toward two techniques, namely zero current switching (ZCS) circuit and the polarity techniques

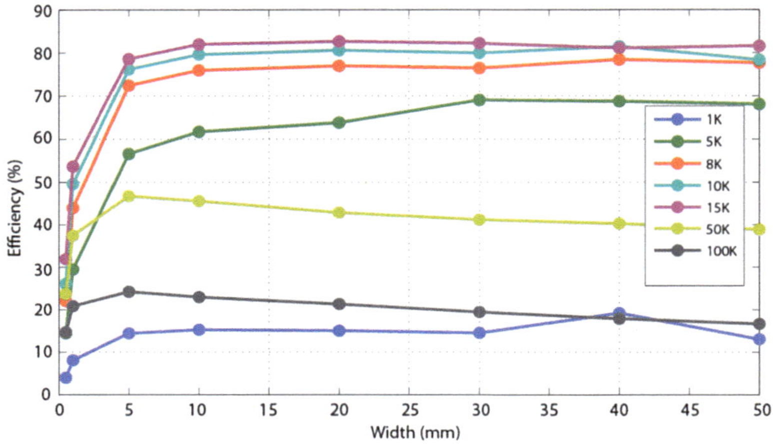

Fig. 3.3 SPICE simulation of the boost converter efficiency with different switches widths and load resistances

as it will be explained in Chaps. 4 and 5, respectively. However, in the following subsections, many techniques are explained briefly for a complete discussion.

3.2.2.1 Zero Current Switching Control Circuit

The ZCS control is an essential circuit for the synchronous inductor-based DC–DC converter. The ZCS is used to turn off the PMOS switch when the inductor current reaches zero. If the PMOS switch is turned off before or after the zero current point, then the efficiency and the output voltage will be affected which results to performance degradation. Further, measuring the inductor current directly is not an easy task especially in low power application. More details about the ZCS control will be explained in Chap. 4 and a detailed example will be presented.

3.2.2.2 Polarity Control for TEG

In many applications, the temperature gradient across the TEG might be flipped such as in the human body harvesting (summer and winter) and indoor and outdoor harvesting applications. Hence, the voltage polarity will flip if the temperature polarity flips. However, the main condition for the inductor-based DC–DC converter to operate is to have a positive input voltage, otherwise, the converter will not be able to work. Hence, a polarity control is necessary to evaluate the input voltage polarity and physically flips the connection of the TEG when needed. More details about the polarity techniques will be explained in Chap. 5 and a detailed example will be presented.

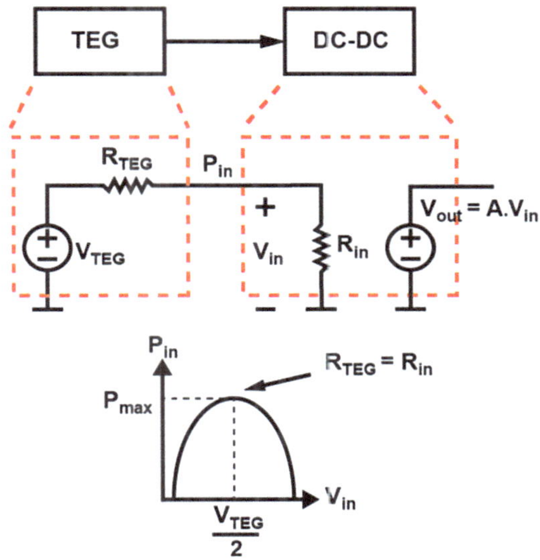

Fig. 3.4 Illustration of MPPT concept for TEG harvester

3.2.2.3 Maximum Power Point Transfer Techniques

The maximum power point transfer (MPPT) technique is used to match the impedance of the inductor-based converter with the impedance of the TEG for maximum power transfer. As shown in Fig. 3.4, the impedance looking into the TEG is resistive (R_{TEG}). As such, the power profile of the TEG is a square function where the peak happens when the load, which is the input impedance of the DC–DC converter, equals to the R_{TEG}. Different techniques are reported in the literature to match the impedance with the TEG. One technique uses a feedback control to force the input node to be half the open circuit voltage by sampling it periodically and dividing it by half using capacitors [68]. This technique is suitable in applications where wide temperature difference is present such that the TEG resistance varies a lot. For applications where the TEG resistance does not change much, then the operating frequency can be set to achieve the maximum power transfer using Eq. (3.5).

3.2.2.4 Startup Circuit

The startup circuit is an important circuit in low voltage applications. It is used to initiate the DC–DC conversion by generating an initial high voltage from the low TEG output voltage. Further, it is needed because of the low output voltage generated from the TEG in some applications such as human body thermal energy. Many techniques have been reported in the literature to kick-off the conversion. Motion activated switch has been used in [27]. The off-chip transformer was presented in [69]. Such transformers are not suitable for wearable applications and large conduction loss is produced. External batteries startup circuit also was proposed in [70] and RF power kick start in [71]. A fully electronic startup circuit has been reported in [72, 73]. It uses an LC oscillator and a voltage multiplier to generate a high voltage. However, the design is sensitive to parasitic capacitances. In contrast, [74] presented simpler approach by using one additional NMOS transistor. However, trimming techniques are used in the post-fabrication process to overcome the variations between different chips due to variations in the fabrication process. In addition, this startup circuit needs tuning with programming time of 3 min. Such post-fabrication process is costly and complicated. Finally, in multi-source energy harvesting systems, one source can be used to kick start the conversion, while another source can be used for continuous harvesting. For example, in [18], an RF energy source is used as to startup a TEG energy harvesting from human body heat.

3.2.3 Power Conversion Architectures

Figure 3.5 summarizes different power conversion architectures for TEG application that are reported in the literature. As depicted in Fig. 3.5a, an inductor-based converter works as the interface circuit to extract the maximum power from the TEG followed by a regulation circuit [18, 27, 66, 75, 76]. This method is efficient because of the usage of the inductor converter. However, the end-to-end efficiency is the multiplication of the efficiency for inductor converter and the regulator. Figure 3.5b presents a technique using a charge pump to integrate the converter

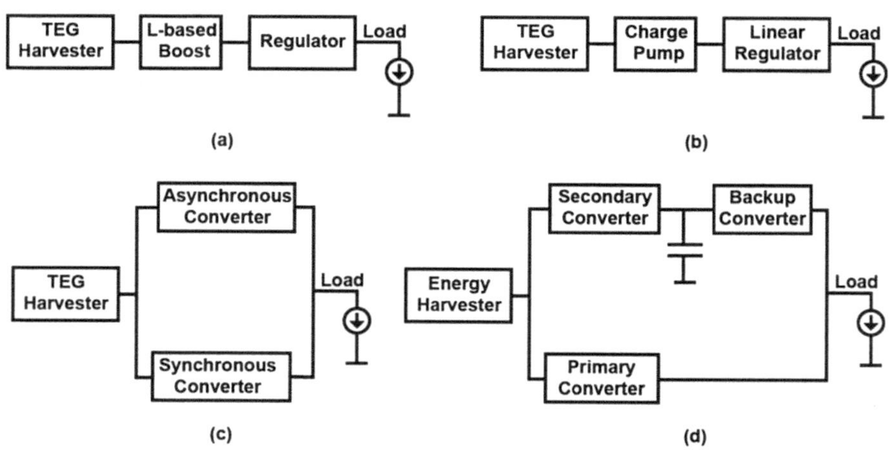

Fig. 3.5 Power conversion architectures. (**a**) Typical architecture for energy harvesting [18, 27, 66, 75, 76]. (**b**) Charge pump-based circuit [77]. (**c**) Multi-step technique using both asynchronous and synchronous converters [72]. (**d**) Dual-path scheme [78]

on-chip compared to the inductor converter [77]. However, the charge pump has lower efficiency compared to the inductor converter. In Fig. 3.5c, it uses a multi-step technique where both asynchronous and synchronous converters are utilized [72]. Finally, a dual-path architecture of two-stage power conversion proposed in [78]. The dual-path scheme uses two parallel converters, one for storing and backup and the other for direct driving the load. Here, during light load conditions, the primary path is activated to drive the load and one DC–DC converter is supplying the current to the load which improves the efficiency of the system. But once the load strives for more current, the other path starts to operate to supply more current, and the system gets similar to the one shown in Fig. 3.5a.

3.2.4 System Robustness

CMOS deep nanometer nodes (65 nm and below) provide the advantages of high speed because of small transistor length with low power consumption due to supply scaling. However, with CMOS node scaling, the process, voltage, and temperature (PVT) variations impact on the system performance increase. As such, the system performance becomes a function of PVT which is undesirable. The threshold voltage variation increases as the length shrinks in the deep node since lower dopant number is available in the channel [79].

To overcome this issue, designers usually overdesign their circuit which results in a lot of design time without ensuring that the circuit maintains the performance due to random variations. Some researchers have started locking into different techniques to solve PVT variations such as Built-in-Self-Test (BiST) and Built-in-Self-Calibration (BiSC) for different applications [80]. This will make the system smart to test itself, check the performance, and then doing some kind of feedback to correct it.

Chapter 4
Zero Crossing Switching Control for L-Based DC–DC Converters

4.1 Background and Prior Work

As explained in Chap. 3, an inductor-based DC–DC converter is widely used in energy harvesting applications due to its high efficiency [67, 76]. The efficiency of the converter depends on the inductor quality factor, the technology used, losses in the converter and the control circuit. In addition, the synchronous inductor-based DC–DC converter is used to reduce the losses associated with the diode in the asynchronous converter [72]. However, the synchronization (using the ZCS control) between the switches in the synchronous converter is essential to maintain a high efficiency [81].

In order to maximize the efficiency of the boost converter, losses in the circuit should be carefully analyzed. Conduction and switching losses of the NMOS, PMOS, and series resistance of the inductor contribute in the overall losses budget. In addition, leakage current through the switches affects the efficiency. Further, the power consumption and the leakage of the control circuitry should be minimized. ZCS control plays a major role in maintaining the efficiency of the inductor-based converter. Figure 4.1 shows the synchronous inductor-based boost converter used for TEG harvester along with the ZCS. As discussed in Chap. 3, the ZCS is used to turn off the PMOS switch when the inductor current reaches zero so that the charge in the output capacitor is preserved.

Different techniques can be implemented to detect the zero current point such as sensing the voltage polarity across the PMOS switch using a comparator-controlled MOSFET (active diode) as shown in Fig. 4.2a [67, 82]. However, this technique requires a very fast comparator to close the PMOS switch once the polarity across the PMOS is flipped. Any delay in the comparator could result in depleting the charge from the output capacitor and degrading efficiency.

Further, in [83], the zero current is detected by forming an inductor current loop using a switch connected in parallel with the inductor as shown in Fig. 4.2b. This switch is activated at the end of the discharging phase so that the polarity of the

M. Alhawari et al., *Energy Harvesting for Self-Powered Wearable Devices*, Analog Circuits and Signal Processing, DOI 10.1007/978-3-319-62578-2_4

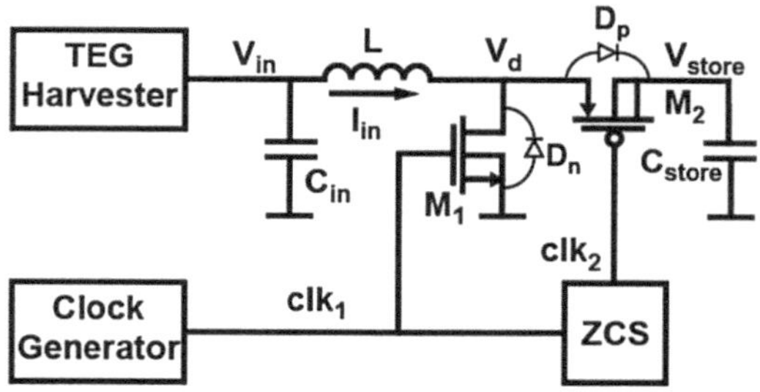

Fig. 4.1 Synchronous inductor-based DC–DC converter with ZCS block

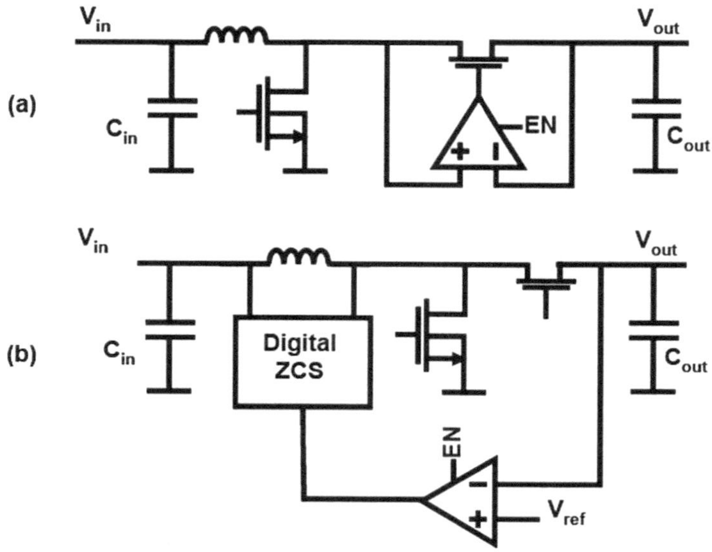

Fig. 4.2 Prior art in zero current detection circuit using (**a**) active diode and (**b**) voltage polarity of the inductor

current is sampled and the zero current point is evaluated. This technique depends on the series resistance of the inductor to form the voltage polarity which could vary from an inductor to another.

Another technique is to monitor the inductor voltage from the high side (V_d) to evaluate the value of the inductor current [75]. In [75], if the PMOS switch is turned off earlier or late than the zero current point, then V_d becomes positive or negative, respectively. As such, the current polarity can be evaluated by adjusting the PMOS turn-on time to approximate the location of the zero current point.

To better understand the early and late ZCS technique, Fig. 4.3 shows an illustration of the boost converter behavior in early and late cases. Inspection of Fig. 4.3b leads into several key observations regarding the discharge path of the inductor current. For early condition, the parasitic diode between the source and bulk of the PMOS is turned on, which means V_d becomes higher than V_{out} by V_{Th_p}.

Fig. 4.3 ZCS concept by monitoring V_d node

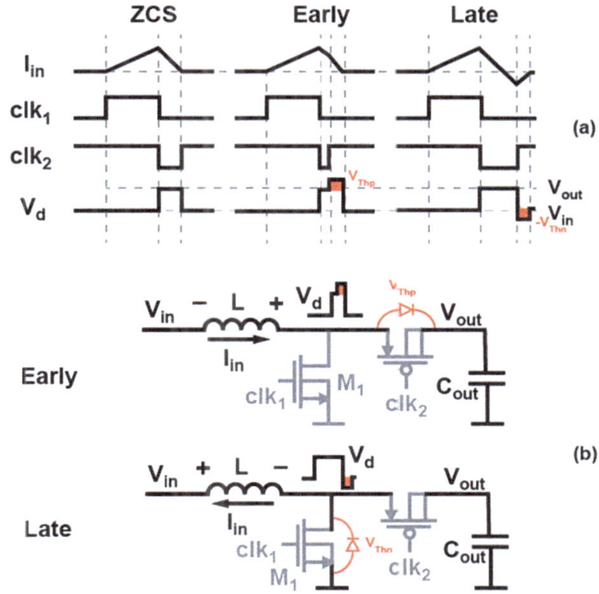

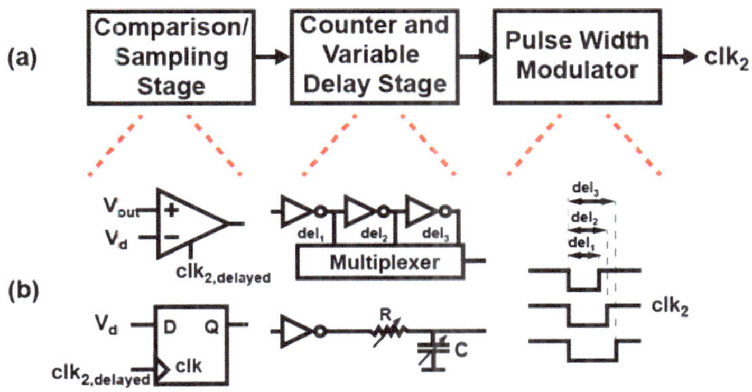

Fig. 4.4 Implementation of ZCS control circuit

In contrast, for the late case, the parasitic diode between the drain and bulk of the NMOS is activated and V_d becomes lower than ground by V_{Th_n}. As such, a control circuit can be designed to evaluate V_d level after the PMOS switch is turned off (at the positive edge of clk_2) such that early and late states can be clearly detected.

Figure 4.4a shows a block diagram of ZCS control circuit based on the concept shown in Fig. 4.3. The first stage is the comparison and sampling stage that is required to sense the level of V_d after turning off the PMOS switch. As depicted in Fig. 4.4b, the detection can be done by comparing V_{out} and V_d after the PMOS is turned off using a clocked comparator activated using a delayed version of clk_2

so that V_d state is detected. In addition, a simple D-FF can be used to sample V_d to detect early and late cases. Once the V_d state is detected, a simple counter and delay elements can be used in order to change the pulse width of clk_2. Delay elements can be constructed by simple inverter chains with multiplexers [27]. Further, capacitor and resistor banks can be used to alter the clk_2 pulse width [66, 84]. Finally, a pulse width modulator is used to generate clk_2 at different delay steps.

There are some challenges when using the V_d method. When the input voltage is changed, then the output voltage will change too when the converter is working in open loop fashion. Therefore, the zero current point will vary with the input voltage level. Hence, the ZCS circuit should cover a wide delay range such that the zero current point can be always detected. To cover a wide delay range, many delay elements should be used which translates to a large area and complex design. Further, the resolution of the delay plays an important role in maintaining the efficiency. As such, the main issue is whether the ZCS circuit should use coarse delays to cover wide delay range which compromises the efficiency, or to use fine delay elements to maintain the efficiency while increasing the chip area as many delay elements will be used.

To explain more, Fig. 4.5 shows SPICE simulation results of an inductor-based DC–DC converter efficiency with ZCS circuit at a load of 10 kΩ. The x-axis represents the time difference between the actual zero current point and the turn-off time of the PMOS switch. At this particular case, the pulse width clk_2 of the PMOS switch at which the zero current is achieved is equal to 20 ns. The negative and the positive time difference T_{diff} represent early and late conditions, respectively. As depicted in the figure, the efficiency drops if the time difference gets larger for both cases. In the late case, a negative inductor current flows from the output capacitor which degrades the efficiency. In contrast, in the early case, the inductor current passes through the parasitic diode of the PMOS switch which causes a loss in the efficiency due to the internal resistance and the voltage drop of the diode. Note that for the early case in Fig. 4.5, the converter becomes asynchronous at T_{diff} less than −520 ns.

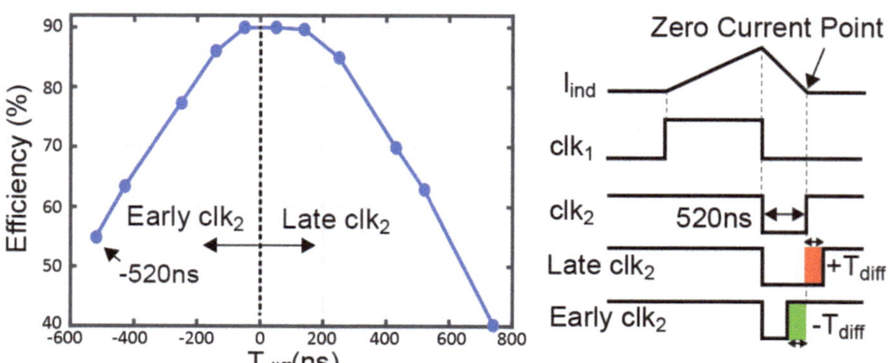

Fig. 4.5 SPICE simulation of the efficiency of boost converter vs. time difference between clk_2 and ZCS point in late condition case

4.2 Example of ZCS Control Circuit

An example of improved ZCS control circuit, compared to [75], is presented in this section where the number of delay steps is increased to cover wide delay range while achieving a fine resolution with few control bits [81]. The work in [81] was proposed by the authors of this book to improve the efficiency and the dynamics of the inductor-based boost converter. Figure 4.6 shows the proposed system that includes the TEG model, an inductor-based DC–DC converter, and the proposed ZCS control. The ZCS consists of the following. A D-FF to detect the status of the V_d voltage, a 3-bit up/down counter to increase/decrease the clk_2 pulse width based on the V_d voltage, a thermometer decoder to translate the binary values of the counter into the corresponding thermometer value, an FSM controls the modes of operation for the overall system, a register to save the counter value of the coarse mode, a multiplexer to change the mode to fine delay, a coarse and fine delay banks that generates the delay of clk_2, and finally the clk_2 generator that generates clk_2 based on the delay from coarse and fine banks.

4.2.1 *Coarse/Fine ZCS Techniques*

To explain the operation of the proposed ZCS, the FSM activates the coarse delay to change the width of the clk_2 pulse so that the zero current point is located. Then the FSM stores the coarse delay value in a register and changes the mode to the fine

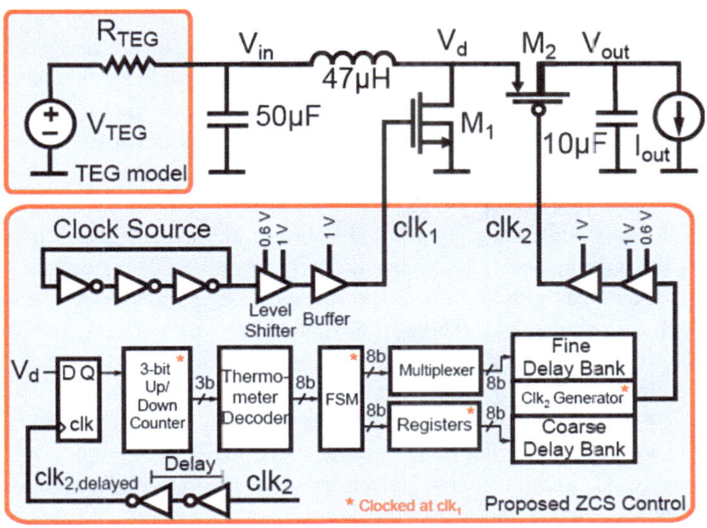

Fig. 4.6 Block diagram of the proposed ZCS circuit

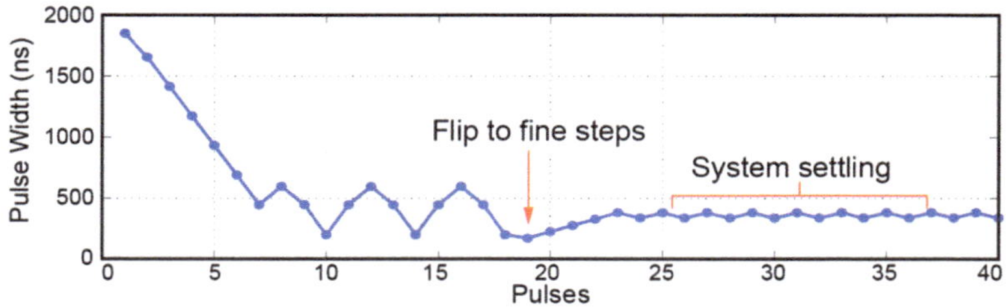

Fig. 4.7 SPICE simulation of system dynamics for coarse and fine delay modes

delay in order to optimize the zero current location. Even if the zero current point is moved due to the input voltage change, the FSM tracks the new zero current point. As such, the proposed system has the flexibility to move from coarse to fine delay or vice versa, seamlessly. Note that the delay to generate $clk_{2,\mathrm{delayed}}$ is defined as the time required for V_d to settle (early or late) which should be set right for proper evaluation. Extra bits from a serial register are added to the design to account for different variations.

The 3-bit counter is shared between the coarse and fine modes so that 56 fine delay steps (8 fine steps in each coarse step which is equal to 7, 8 minus 1) are achieved with only 3-bit design. As such, the system has an effective number of bits equal to 5.8 bits ($ln56$). Note that two external power supplies are used. The digital part is operated at 0.6 V to reduce the power consumption and the leakage. Further, the power switches are driven by 1 V (using two level shifters and buffers) such that their losses are minimized. The power consumption for both 0.6 and 1 V is included in the overall efficiency measurements.

Figure 4.7 shows SPICE simulations of the system dynamics when ZCS is operated. As shown in Fig. 4.7, the coarse delay searches for the zero current point and toggles around it. Then the FSM switches to the fine mode for final settling. The coarse and fine steps are 250 ns and 50 ns, respectively. Note that the settling time of the proposed ZCS control is faster than a fixed delay system for the same effective number of bits. To explain, a system that has fixed delays with the same resolution of the proposed system (56 delay steps) needs 5.8 bits ($2^{5.8}$). Even if the zero current location is in the middle range, the fixed delay system needs 28 cycles (56/2) to reach the zero current point. In contrast, the proposed system needs a *maximum* of seven coarse cycles plus eight fine cycles and extra cycles for the transition to cover the whole delay range. There are eight extra cycles during the startup (as shown in Fig. 4.7) which are reduced to one cycle during normal operation as will be explained in the following section.

Figure 4.8a shows the circuit that generates the delay for clk_2 pulse. In addition, Fig. 4.8c, d shows the operation of the circuit. As depicted in the figure, when clk_1 is high, M_3 charges C_1 and the coarse capacitors since cnt_1 is initially set to logic high which makes sig_1 discharged to ground. As such, a maximum delay is generated. Once the evaluation of V_d is detected to be late, cnt_1 bits start to turn off coarse

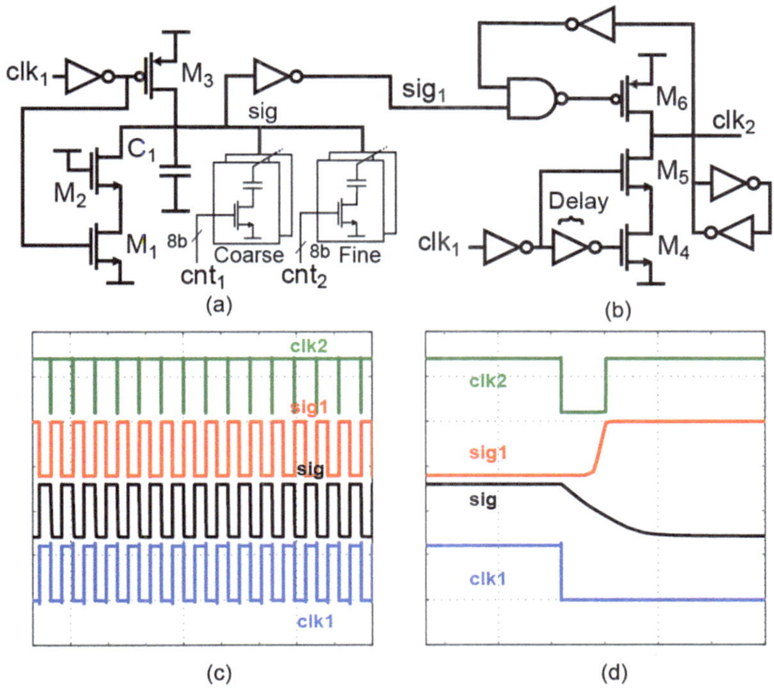

Fig. 4.8 (**a**) Coarse and fine delay circuit. (**b**) clk_2 generation circuit. (**c**) Simulation shows the operation of the circuit. (**d**) Zoom-in version of **c**

capacitors one by one till early evaluation of node V_d is detected. As such, the FSM waits for toggling between late and early cases in coarse condition for eight clock cycles then shifts the state to fine tuning delay for final system settling.

Figure 4.8b shows the pull-push circuit that generates clk_2 signal. As illustrated in the figure, sig_1 signal from Fig. 4.8a is an input to the circuit in Fig. 4.8b. When clk_1 is high, sig_1 is low and clk_2 is high. In contrast, when clk_1 goes to ground, clk_2 is discharged through M_4 and M_5 and its state is held by the output latch. The state of clk_2 stays the same till sig_1 signal changes its state depending on the coarse and fine delay banks. Once sig_1 goes high, then clk_2 is pulled high through M_6. Note that M_4 and M_5 can be both on at the same time due to the delay introduced at M_4 gate. Finally, Fig. 4.8c displays the coarse and fine delay banks. The banks consist of switches and capacitors to introduce delay depending on the control signals cnt_1 and cnt_2.

To generate the system clock, a ring oscillator is utilized as shown in Fig. 4.9. As depicted in the figure, tuning capacitors are used to adjust the frequency due to the process variations. Note that the *control* signal is generated from a serial register.

Figure 4.10 shows the SPICE simulation results of the overall efficiency and output power at V_{TEG} of 50 mV and R_{load} of 15 kΩ connected to V_{store}. As depicted

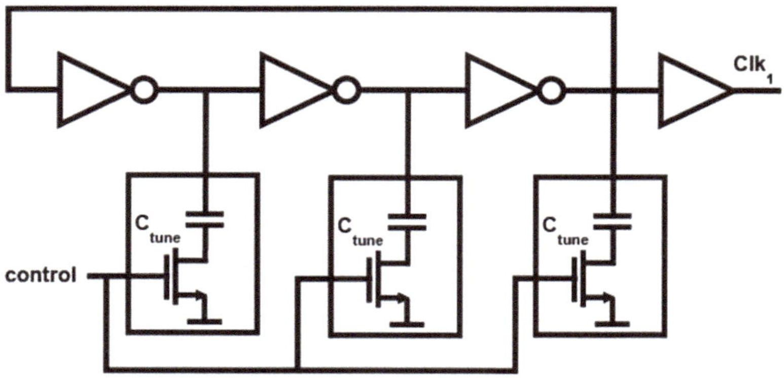

Fig. 4.9 Ring oscillator to generate the system clock with tuning capacitors

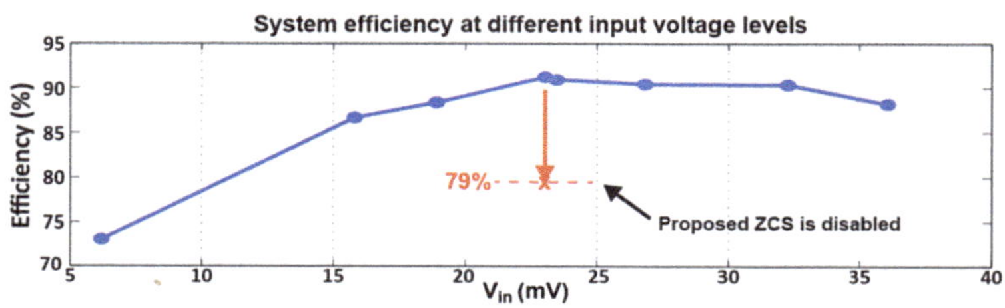

Fig. 4.10 Efficiency, output power, and output voltage V_{store} at different V_{in}

in Fig. 4.10, more than 85% efficiency can be achieved where 90% efficiency is obtained at the maximum power transfer point. In contrast, to confirm the advantage of the proposed ZCS circuit, the efficiency of the system is calculated when the proposed ZCS is disabled and only the coarse resolution is activated at the maximum power transfer point in Fig. 4.10. At that point, approximately 79% efficiency is achieved compared to ∼90% in the case of the proposed ZCS of using fine resolution. This confirms the improvement in efficiency of more than 10% when using the proposed ZCS circuit. In addition, a relative efficiency could be used to show the improvement of the proposed ZCS circuit which is equal to $90/79 - 1 = 14\%$.

4.3 Measurement Results

Figure 4.11 shows the measurement setup of the overall system that was discussed in Fig. 4.6. A DC power analyzer is used to emulate the voltage generated from a TEG which has a series resistance of 7 Ω (to emulate G2-30-0313 from Tellurex). Further, the DC power analyzer is used to measure the efficiency of the DC–DC converter.

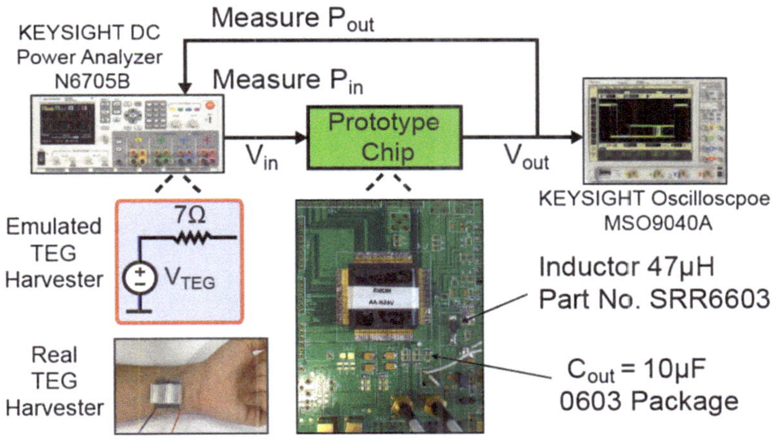

Fig. 4.11 Measurement setup of overall system

In addition, an oscilloscope is utilized to measure the time domain signals. Finally, a real TEG (G2-30-0313 from Tellurex) is placed on a human hand for real case experiment instead of the emulated one.

The connection between the off-chip components (i.e., inductor, capacitors) and the chip contributes in the overall loss of the system due to parasitic resistances and capacitances. As such, to reduce these losses from the printed circuit board (PCB), the copper width is increased, the connectors are placed as close as possible to the pin of the chip and ground planes are designed to reduce the resistance loss.

Figure 4.12a shows the measured efficiency at different loads and TEG voltages. The efficiency is an important metric to show the percentage of the delivered output power relative to the losses in the circuit. The efficiency is defined as the output power P_{out} divided by the input power from the TEG, P_{in} at the input voltage V_{in} which will be equal to half the V_{TEG} since clk_1 frequency is chosen to achieve the maximum power transfer. The peak efficiency reaches 81% at V_{TEG} of 50 mV. In addition, Fig. 4.12b shows the output voltage at different V_{TEG} voltages and load currents which reflect the gain of the converter and the ability to support the output voltage for the load. Note that the efficiency measurement includes all the losses in the system (conduction losses, switching losses, leakage, quiescent current, etc).

Figure 4.12c shows the relationship between the input and the output voltage. As indicated in the figure, the output voltage is larger than 1 V over the range of 20–50 mV. Note that the output voltage is saturated when the input voltage is larger than 50 mV where no regulation is implemented in the converter. The reason is that the ZCS control circuit is saturated and not able to provide more delay steps for input voltages larger than 50 mV as will be discussed in Fig. 4.14.

Inspection of Fig. 4.12 leads to a key observation regarding the DC–DC converter performance. First, the high efficiency is achieved mainly because of the usage of high-quality factor inductor with an advanced 65 nm CMOS and low converter

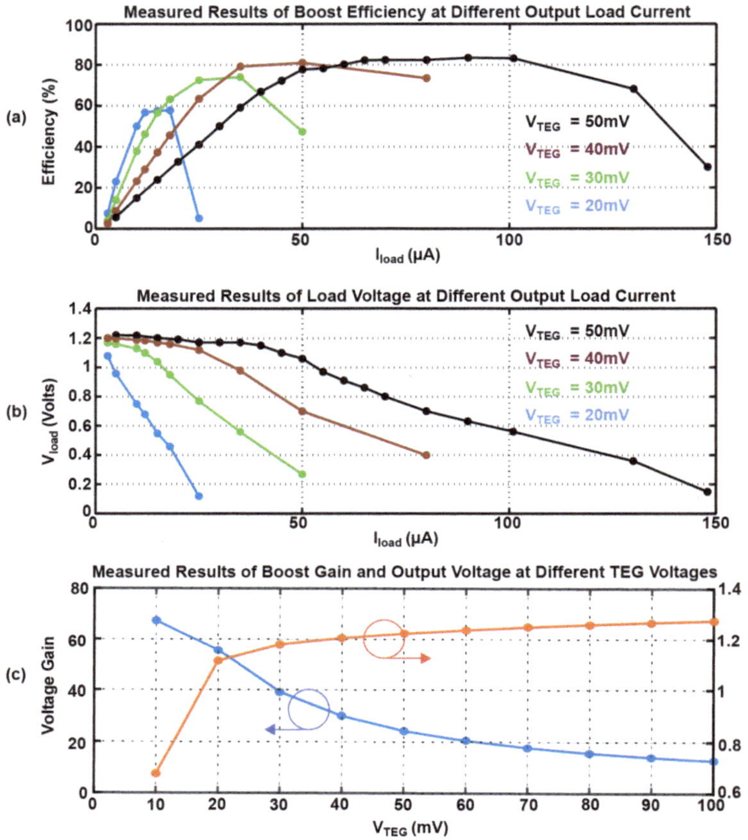

Fig. 4.12 Overall system with inductor-based converter and the proposed ZCS

losses. However, the proposed ZCS control circuit helps in maintaining the high efficiency as was discussed in Fig. 4.5. Note that this system is part of a biomedical SoC which requires an output voltage from the DC–DC converter to be between 0.9 and 1.3 V [85]. However, Fig. 4.12 shows the performance of the DC–DC converter over a wide range of output currents and voltages for complete characterization.

The measured time domain results of the system in Fig. 4.6 at no load and V_{TEG} of 50 mV are presented in Fig. 4.13. This graph is essential to confirm the operation of the early/late conditions of the proposed ZCS. As indicated in Fig. 4.13a, the inductor current ramps up during the on-time of clk_1 and then ramps down during the on-time of clk_2. The inductor current is measured by placing a 2.2 Ω in series with the inductor and then the voltage across the resistor is measured. Since the voltage across the resistor is low (∼18 mV), the noise from the oscilloscope affects the measurement. As such, a digital filter inside the oscilloscope is used to filter the noise on the inductor current waveform.

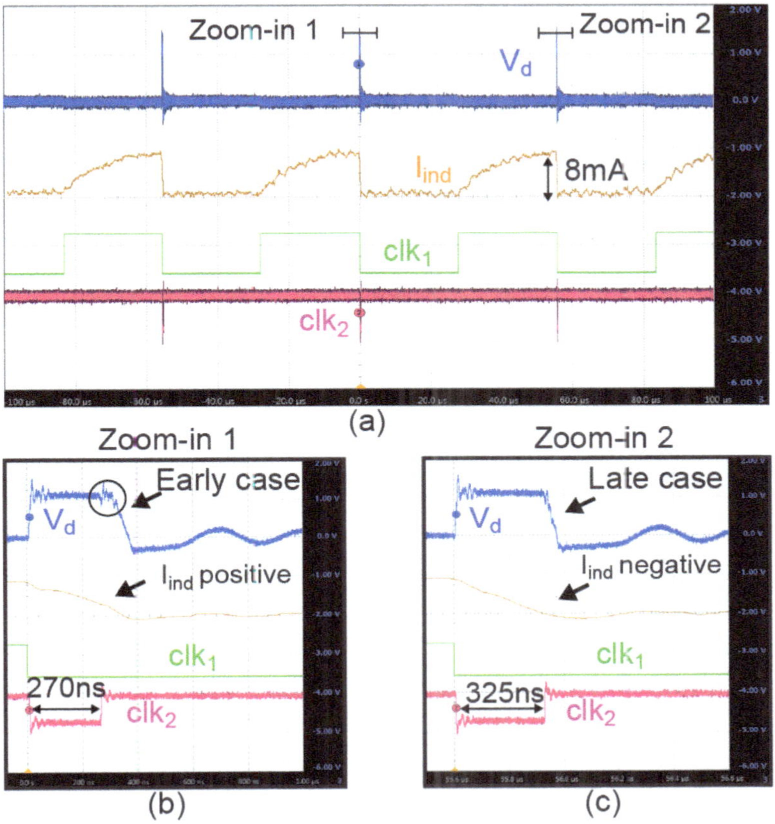

Fig. 4.13 (**a**) Measured time domain result of the system in Fig. 4.6. (**b**) Early condition. (**c**) Late condition

Figure 4.13b, c shows the early and late conditions of the inductor current, respectively. As indicated in Fig. 4.13b, the early condition of the inductor current is detected which pumps the V_d node. In contrast, Fig. 4.13c shows the late condition of the inductor current. Note that the pulse width of clk_2 is changed from 270 to 325 ns (55 ns difference) which represents the resolution of the fine delay. Finally, the 2.2 Ω resistor is inserted only in this setup to measure the current and it is removed in other measurements since it affects the output voltage level. Finally, a negative inductor current is noticed in the early condition which should not be confused with the late condition. This current is not from the output capacitor but from a transient effect that happens due to the formation of a resonance loop between the inductor, the parasitic diode of the NMOS switch, and the input capacitance.

Figure 4.14a shows a time domain result when the TEG voltage changes from 20 to 100 mV. This graph is important to highlight the dynamics of the proposed ZCS

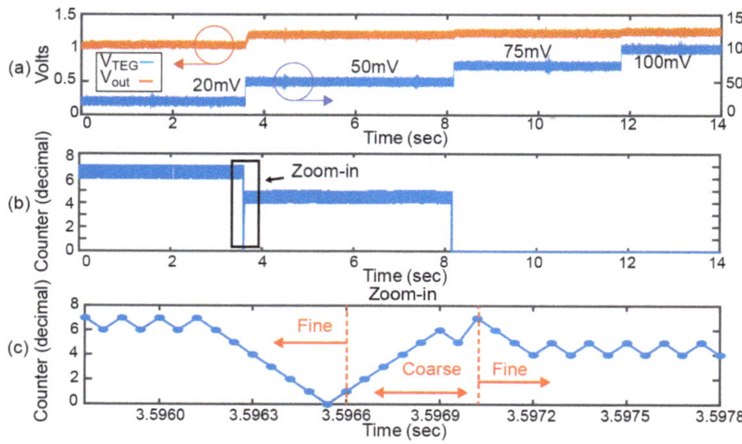

Fig. 4.14 (**a**) Time domain output voltage at different TEG voltages. (**b**) Counter value that represents the coarse and fine dynamics. (**c**) Zoom-in version of graph **b**

control and to confirm the operation of the system. Figure 4.14b shows the value of the 3-bit counter, which is shared between both the coarse and fine delay steps. Initially, the TEG voltage is 20 mV and the fine delay steps toggle between 6 and 7. Once the TEG voltage becomes 50 mV, then the FSM goes into the following states as shown in Fig. 4.14c. First, the fine steps are changed to search for the new zero current point. Once the fine steps are saturated, then the FSM changes to coarse steps. The coarse steps are changed till the zero current point is found between the value 5 and 6. Then, the mode is changed to fine steps for final settling between 4 and 5. One should note that it needs one clock cycle to change the mode from coarse to fine steps. Further, the counter value for input voltage larger than 50 mV is zero. This means that the ZCS circuit is not able to provide more delay steps outside the input voltage range (20–50 mV) and the zero current point is not detected efficiently. This is why we see saturation in the output voltage for TEG voltages larger than 50 mV as shown in Figs. 4.12c and 4.14a. However, our range of interest is from 20 to 50 mV.

Figure 4.15 shows the measured TEG voltage and output voltage for the experiment of real TEG which is placed on a human wrist at room temperature as was shown in Fig. 4.11. As depicted in the figure, the TEG input voltage settles at 30 mV and produces a no-load output voltage of 1.1 V. These results confirm the feasibility of using a real TEG placed on the human body to generate enough load voltage to power a low power load.

The die photo is shown in Fig. 4.16a. The total active area is less than 0.04 mm^2 with 0.3× increase in area due to the proposed ZCS control that includes the FSM and the fine delay elements. The layout in Fig. 4.16b shows different parts of the chip. To compare the system performance with recent work, Table 4.1 shows that the proposed system achieves high efficiency at low output power at a range of 20–50 mV and a large number of steps with few control bits.

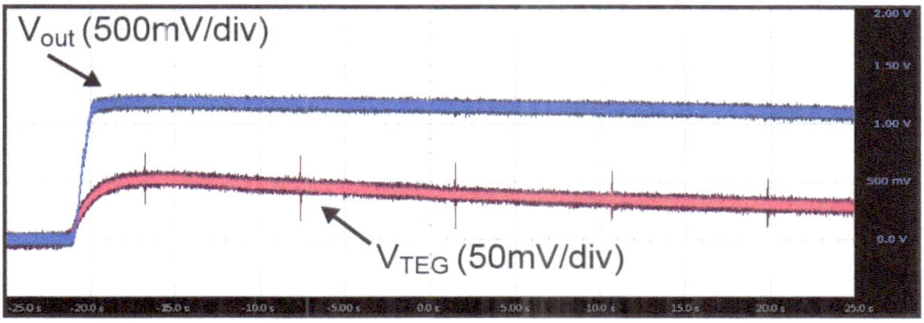

Fig. 4.15 Measured input and output voltage with real TEG harvester

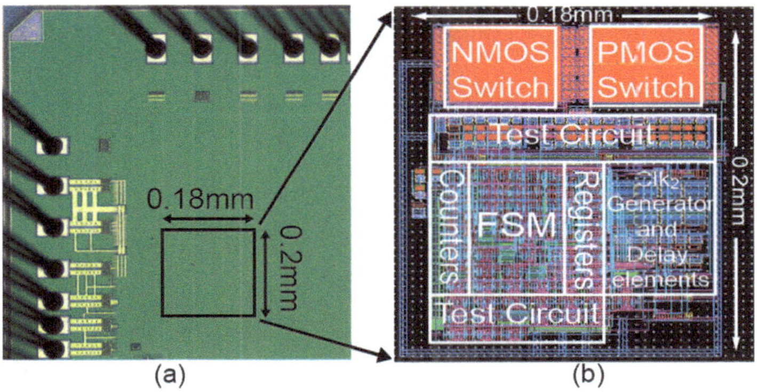

Fig. 4.16 (**a**) Die photo of the prototype chip. (**b**) Layout showing different circuit parts

Table 4.1 Comparison table with state of the art

Parameter	Yogesh JSSCŠ11	Weng JSSCŠ13	Shriva JSSCŠ15	Im ISSCCŠ12	Damodaran TPELŠ15	Chen TCAS-1Š15	Carlson JSSCŠ10	Work in this book
Harvesting	TEG	TEG	TEG/Solar	TEG	Solar	TEG	TEG	TEG
Technology	0.35-μm CMOS	65-nm CMOS	130-nm CMOS	130-nm CMOS	180-nm CMOS	180-nm CMOS	130-nm CMOS	65-nm CMOS
Minimum input voltage	25mV	50mV	10mV	40mV	270mV	100mV	20mV	20mV
Maximum output power	280μW	1.1mW	22mW	2.7mW	100μW	1mW	100μW	55μW
Maximum output voltage	1.8V	1.2V	1.1V	2V	1.8V	0.5V	1.0V	1.2V
Zero crossing technique	V_d method	V_d method	Latched comparator	Diode (asynchronous)	Latched comparator	Latched comparator	V_d method	V_d method
Control bits/delay steps	3-bits/8 steps	4-bits/16 steps	NA	NA	NA	NA	4-bits/16 steps	3-bits/56 steps
Efficiency at 50mV Input	58% at 50μW	63% at 100μW	68% not reported	33% not reported	NA	NA	65% at 40μW	81% at 55μW
Peak efficiency	58%	73%	83%[a]	61%	86%[b]	83.4%[c]	75%	81%[d]

[a]At 300 mV input [b]At 500 μW output power [c]At 100 mV input and 500 μW output power [d]External 0.6 and 1 V are used

Chapter 5
Polarity Mechanism for Thermoelectric Harvester

5.1 Prior Work in TEG Polarity Mechanism

The output voltage polarity from the TEG is related to the temperature difference polarity across the TEG as shown in Eq. (5.1), where T_1 and T_2 are the temperature of the first and the second side of the TEG, respectively [27]. From Eq. (5.1), if T_1 is higher than T_2, then V_{TEG} is positive. To explain, Fig. 5.1a, b shows a diagram of a TEG under normal and reversed temperature polarity. As indicated in Fig. 5.1a, T_1 is higher than T_2, which results in a positive output voltage, V_{load}. In contrast, Fig. 5.1b shows the case when T_2 is higher than T_1, which results in a reversed output voltage, V_{load}.

In many applications, the polarity of the TEG is either changing or unknown. For example, thermal harvesting from the human body depends on the difference between the body temperature and the ambient temperature [27]. Since the ambient temperature could be higher or lower than the body temperature, the TEG output voltage could be positive or negative. Further, a TEG can be useful with changing environmental conditions such as indoor and outdoor use [86]. The power conversion circuit has to see a positive voltage from the TEG to work properly. Hence, a polarity controller is necessary to detect and reverse the TEG terminals when it is negative so that the harvesting is not interrupted.

$$V_{TEG} = S_e(T_1 - T_2) \tag{5.1}$$

Figure 5.2 shows a typical block diagram for an auto-polarity structure in thermal harvesting systems. As indicated in the figure, the polarity controller monitors the polarity of the TEG voltage. The switch matrix acts as the interface circuit to the inductor-based converter. If a negative polarity is detected, then the polarity controller sends a signal to the switch matrix to flip the connection such that V_{in} is always positive. The switch matrix circuit needs to have large switches to provide a negligible resistance compared to the TEG resistance. The input capacitance (C_{in})

M. Alhawari et al., *Energy Harvesting for Self-Powered Wearable Devices*, Analog Circuits and Signal Processing, DOI 10.1007/978-3-319-62578-2_5

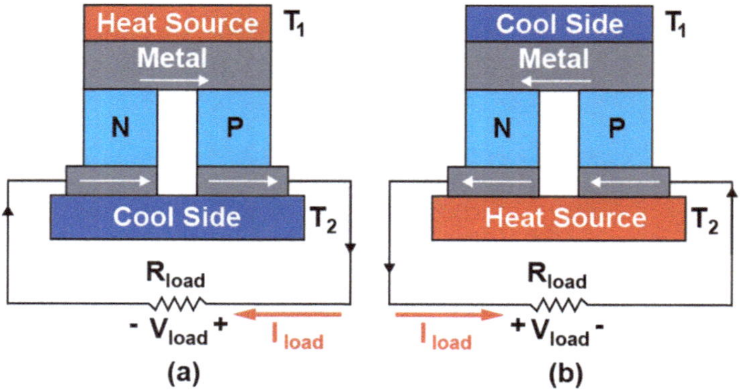

Fig. 5.1 A conceptual diagram of a TEG at (**a**) normal polarity and (**b**) reversed polarity

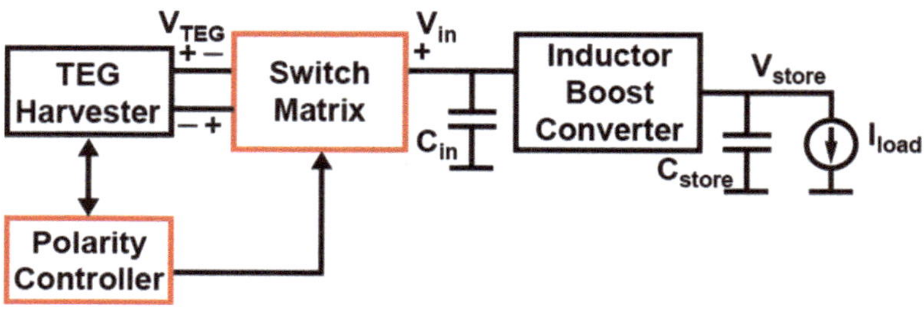

Fig. 5.2 A typical block diagram of a TEG energy harvesting system utilizing a polarity controller and a switch matrix

needs to be large for two reasons. First, the charge injection effect from the switch matrix will be negligible. Second, it will prevent V_{in} to drop during switching.

To highlight the key challenges toward an efficient polarity controller, we consider the classical implementation shown in Fig. 5.3a. Here, a comparator circuit is utilized to compare directly the polarity of the V_{TEG} voltage [86]. Then, a switch matrix is used to adjust the polarity based on the comparator output signal so that V_{in} becomes positive. This technique is useful when the TEG voltage is large, however, in human body thermal harvesting, it requires a careful design for the comparator due to the low TEG voltage (20–50 mV). At low voltage range and due to the process, voltage, and temperature (PVT) variations, designing a comparator to directly detect small voltage change is indeed elusive.

Figure 5.3b shows another approach for detecting TEG polarity where two inductor-based boost converters are connected in parallel, which can be used for power conversion at both input polarity cases [87]. As depicted in the figure, the input voltage of each converter is reversed such that one converter works when the polarity is positive while the other one works when the polarity is negative. This approach requires two inductor converters which increase the cost and the

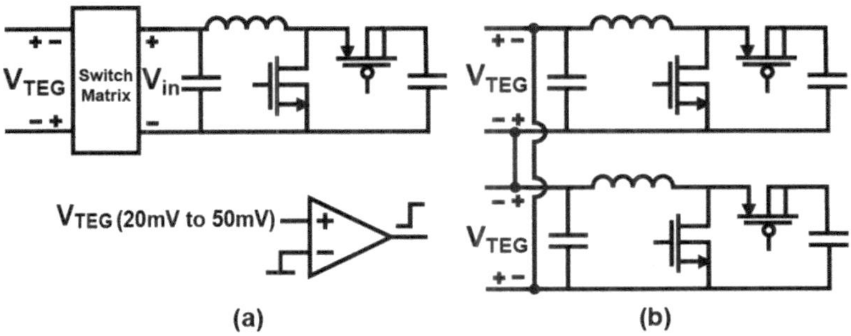

Fig. 5.3 Prior circuit techniques in monitoring and controlling the polarity of the input voltage. (**a**) Comparing the input voltage directly using a comparator [86]. (**b**) Dual converters method with the opposite input voltage polarity on each [87]

form factor of the system. Hence, this solution is obviously very bulky and not suitable for wearable applications. Since the polarity detection depends on duplicate converters, they are subjected to process and mismatch variations which could affect the system sensitivity and functionality between the two converters. Therefore, a better solution is necessary to achieve an efficient polarity detection while having low power consumption and small overhead.

5.2 Example of Auto-Polarity Control Circuit

In this section, an example of an efficient auto-polarity circuit is presented which was designed by the authors of this book. Figure 5.4 illustrates the architecture of this method [88]. Rather than directly comparing the voltage of the TEG, the proposed technique uses the inductor voltage V_d to evaluate the input polarity in a closed loop system. When a negative polarity is detected, then a switch matrix circuit is used to physically flip the connection of the TEG harvester so that V_{in} becomes always positive. The system is an all-digital, fully integrated that uses only one inductor compared to the dual converters method in [87]. The polarity monitoring is done using the inductor voltage V_d, as it will be shown in the next section, which is much larger in amplitude than the input voltage V_{TEG} during the polarity evaluation. As such, the design of the comparator is relaxed in the proposed system compared to direct comparison method in [86]. Finally, the system has a small area overhead, power consumption, and high efficiency as it will be shown in the following sections.

Figure 5.5 shows the details of the auto-polarity technique. As indicated in the figure, a TEG harvester is connected to an inductor boost converter at the two polarity conditions of the TEG voltage. To evaluate the polarity of the TEG voltage, the PMOS switch M_2 is turned off so that the charge stored in C_{store} is not disturbed. A signal S is used to mask the clk_2 signal during the evaluation cycle. Figure 5.5a,

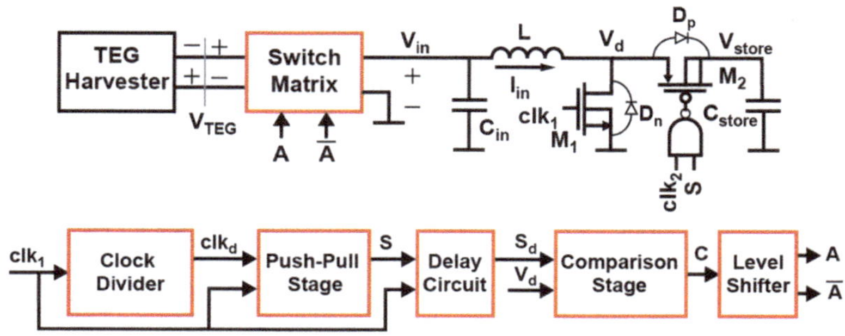

Fig. 5.4 A block diagram of the proposed auto-polarity technique in TEG harvesting systems

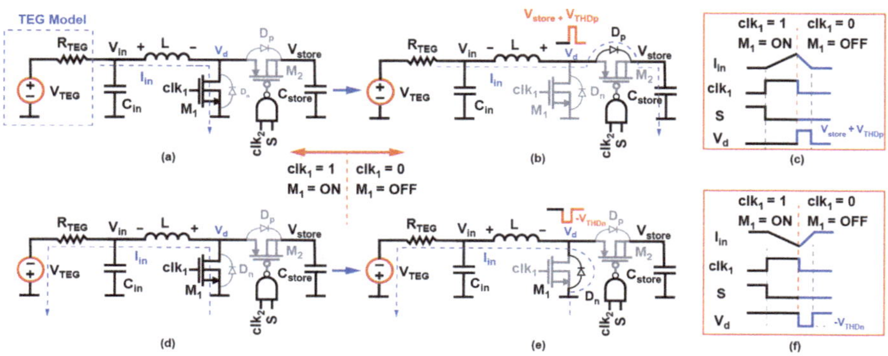

Fig. 5.5 The inductor converter behavior at positive input voltage (**a**) when clk_1 is high and the inductor is charging, (**b**) when clk_1 is low and V_d is raised to $V_{store} + V_{THD_p}$. (**c**) Summary of time domain signals at positive input voltages. The inductor behavior at negative input voltage (**d**) when clk_1 is high and the inductor is charging, (**e**) when clk_1 is low and V_d becomes to $-V_{THD_n}$. (**f**) Summary of time domain signals at negative input voltages

b displays the inductor converter behavior when the input voltage is positive. In Fig. 5.5a, S is low, clk_1 is high, and the inductor current I_{in} is ramped up through M_1. When clk_1 becomes low, as in Fig. 5.5b, then the inductor will force V_d to become $(V_{store} + V_{THD_p})$, where V_{THD_p} is the threshold voltage of the parasitic diode D_p. Thus, the inductor current flows to C_{store} through D_p and the state of V_d at the positive TEG voltage is $(V_{store} + V_{THD_p})$ after the falling edge of clk_1 as illustrated in Fig. 5.5c.

In contrast, Fig. 5.5d shows the inductor converter behavior when the TEG voltage is negative. As depicted in Fig. 5.5d, when clk_1 is high, the inductor is charged in the reversed direction. Once clk_1 becomes low as in Fig. 5.5e, then the inductor will force V_d to be equal to $-V_{THD_n}$. As such, D_n is turned on and the inductor energy is discharged. Hence, the state of V_d at negative TEG voltage is $(-V_{THD_n})$ after the falling edge of clk_1 as illustrated in Fig. 5.5f. Note that the inductor charging loop in Fig. 5.5a, d is the same except the direction of the inductor current.

Thus, the polarity of a TEG can be detected by monitoring the polarity of V_d voltage after clk_1 goes low while M_2 is masked. One should note that the proposed technique is a simple yet effective method which does not directly depend on the absolute value of the TEG voltage. However, it relies on the V_d voltage and the threshold voltage of the parasitic diodes, which is typically larger than 250 mV. Note that during the positive polarity case, V_d amplitude depends on $V_{store} + V_{THD_p}$. Since V_{store} depends on the gain of the converter, then V_d amplitude in the positive case depends on the gain as well. In the presented system, V_{store} can reach up to 1.2 V which makes V_d amplitude in the positive polarity becomes larger than 1.4 V. This helps in relaxing the performance requirements of the comparator used to evaluate V_d. Even if the gain of the converter is small, V_d has a minimum voltage margin of a threshold voltage of the parasitic diodes between the positive and the negative cases. Hence, the polarity evaluation can be performed efficiently.

To determine how frequent the polarity should be evaluated, different trade-offs are investigated in the design. First, if the system has a fast response to the polarity flip, then the system will activate D_p more often, which leads to an efficiency loss. In contrast, if the system has a slow response to the polarity flip, then the system will operate in the wrong polarity for a longer time. This also leads to an efficiency loss due to depleting C_{store}.

The energy loss due to a parasitic diode can be estimated by calculating the inductor current. Assuming an ideal inductor, an input source, and an NMOS switch, the peak inductor current can be calculated from Eq. (5.2) [89]. Where L is the inductance and T_{on} is the on-time of the NMOS. Using the values shown in Table 5.1, V_{TEG} voltage equals to 50 mV, and T_{on} equals to 1/(2×18 kHz), then the peak inductor current is equal to 30 mA. Note that the switching time of the NMOS is much larger than the switching time of the PMOS (few hundreds of ns)

$$I_{ind} = \frac{V_{TEG}}{L} \times T_{on} \tag{5.2}$$

Equation (5.3) shows the energy loss in D_p relative to the stored energy in the inductor. Where R_{diode} is the series resistance of D_p, and $T_{diode,on}$ is the turn-on time of D_p. Ignoring R_{diode}, V_{THD_p} is ~300 mV and $T_{diode,on}$ is ~400 ns (which is extracted from measurement results in Fig. 5.15d), then the percentage of the energy loss per cycle is equal to 17%. If the polarity is evaluated every 16 cycles, then the energy loss is ~1%.

$$\frac{E_{diode,loss}}{E_{ind,stored}} = \frac{(V_{THD_p} \times I_{ind} + I_{ind}^2 \times R_{diode}) \times T_{diode,on}}{0.5 \times L \times I_{ind}^2} \tag{5.3}$$

Table 5.1 A summary table of different parameters used in the boost converter design

Parameter	TEG Voltage (V_{TEG})	Switching Frequency (f_{clk1})	Duty Cycle (D)	Inductor (L)	Input Capacitor (C_{in})	Storage Capacitor (C_{store})
Value	20 mV–50 mV	18 kHz	50%	47 μH	47 μF	10 μF

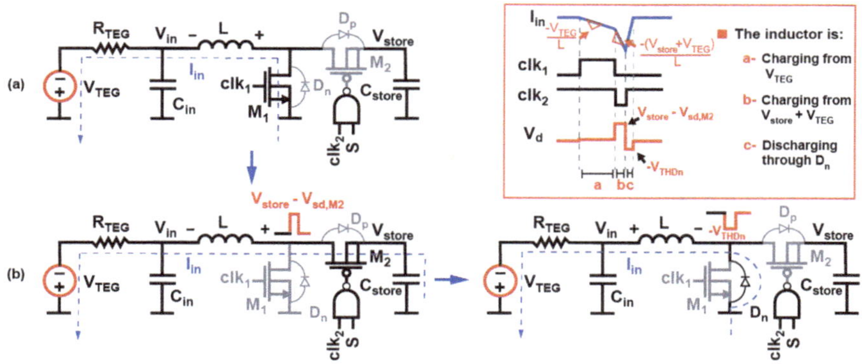

Fig. 5.6 The inductor converter behavior at negative input voltage (**a**) charging the inductor from the TEG, (**b**) charging the inductor from C_{store}, (**c**) discharging the inductor through D_n, and (**d**) illustration of the charging cycles of the inductor

Figure 5.6 illustrates the dynamics of the inductor converter during the reversed polarity case before detected by the proposed polarity method. As depicted in Fig. 5.6a, when clk_1 is high, then the inductor is charged in the reversed direction. Furthermore, as indicated in Fig. 5.6b, on the falling edge of clk_1, clk_2 goes low as well in order to turn on M_2. Hence, the inductor is charged from C_{store} and the TEG voltage as well. This will make V_d voltage less than the output voltage by the voltage drop of the PMOS. Note that, as shown in Fig. 5.6d, the absolute value of the inductor current slope in Fig. 5.6b ($|-(V_{\text{store}}+V_{\text{TEG}})/L|$) is larger than the one in Fig. 5.6a ($|-V_{\text{TEG}}/L|$). The reason is that in Fig. 5.6a the inductor is charged from V_{TEG} while in Fig. 5.6b the inductor is charged from V_{store} in addition to V_{TEG}. However, the time duration at which the inductor is charged from V_{store} (clk_2 pulse width) is much smaller than the TEG voltage time duration (clk_1 pulse width). Finally, as shown in Fig. 5.6c, the inductor is discharged through D_n.

Now, the energy loss per cycle from C_{store} due to the reversed polarity can be calculated. The inductor current from C_{store} can be calculated from Eq. (5.2). For V_{store} equals to 1.2 V, T_{on} is ~400 ns, then the peak inductor current is equal to ~10 mA. Thus, the energy stored in the inductor, which is the same as the energy loss from C_{store}, is equal to 2.3 nJ. If the polarity is evaluated every 16 cycles, then the reversed polarity stays for 15 cycles. This results in a total energy loss from the C_{store} of 34.5 nJ. The energy stored in a 10 μF capacitor at 1.2 V is equal to 7.2 μJ. As such, the energy loss percentage from the output capacitor is less than 1%. Note that in the reversed polarity case, it is important to select a relatively large output capacitance so that only a small percentage of the stored energy is depleted. Therefore, there is a trade-off in selecting the evaluation cycle for the polarity detection, slow versus fast response. In the proposed system, the polarity is evaluated every 16 cycles of the switching frequency clk_1.

To protect the boost converter from the voltage spikes that might happen due to the switching activities, 3.3 V thick oxide transistors are utilized for the NMOS,

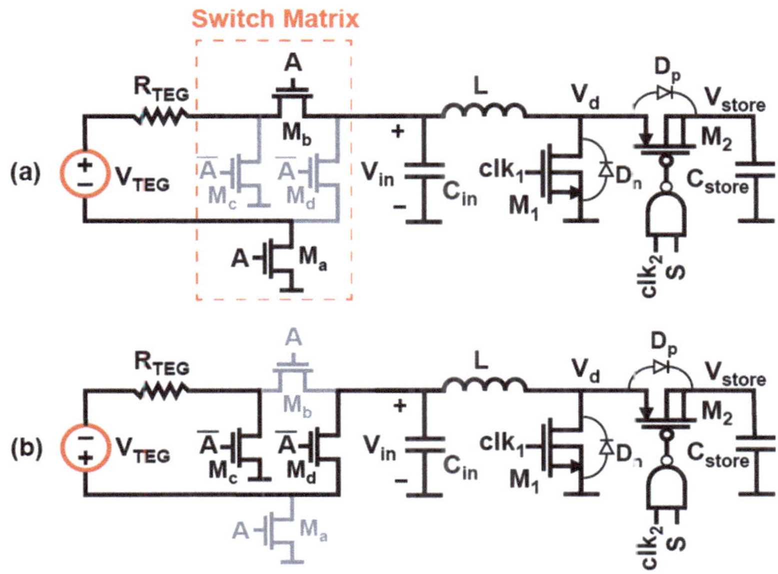

Fig. 5.7 The circuit implementation of the switch matrix when the TEG voltage polarity is (**a**) positive and (**b**) negative

PMOS, the four transistors in the switch matrix as well as the circuits that evaluate the V_d voltage. In addition, an electrostatic discharge (ESD) protection circuit is designed to protect the V_{in}, V_d, and V_{store} if the voltage exceeds 3.3 V.

We now present the circuit implementation of the proposed auto-polarity system that was shown in Fig. 5.4. The switch matrix circuit is presented in Fig. 5.7. The switch matrix consists of four NMOS transistors. They are connected such that the TEG terminals can be physically flipped to maintain a positive V_{in} based on the control signals A and $\bar{A}$. As indicated in Fig. 5.7a, when V_{TEG} is positive, then A is set high to turn on M_a, M_b while keeping M_c, M_d turned off. As a result, V_{in} is positive at the input terminals of the inductor converter. In contrast, Fig. 5.7b shows the case when the V_{TEG} is negative. In this case, A becomes low to turn on M_c, M_d and to turn off M_a, M_b. Hence, V_{in} is maintained positive at the input terminals of the inductor converter.

To explain the effect of the dead time between A and $\bar{A}$ on the inductor converter, we consider the switch matrix circuit in Fig. 5.7. First, $\bar{A}$ is assumed to reach the switch matrix before A (the effect is similar if A is assumed instead). From Fig. 5.7a, since $\bar{A}$ will reach before A, there is a short time where all the transistors in the switch matrix are ON. This means that the TEG source, as well as C_{in}, is shorted to ground for a short period of time. Since C_{in} is large, the loss in the charge is negligible and it will be recovered from the TEG source once A reaches the switch matrix. In contrast, as in Fig. 5.7b, if $\bar{A}$ reaches the switch matrix first, then all the transistors in the switch matrix will be OFF and there is no impact in this case. Note

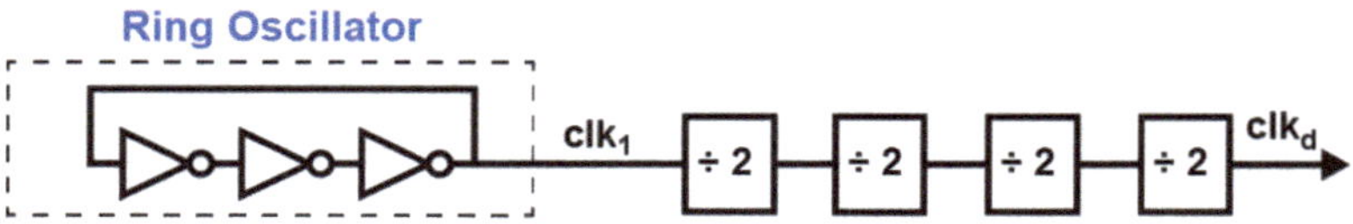

Fig. 5.8 The block diagram of the divide-by-16 circuit

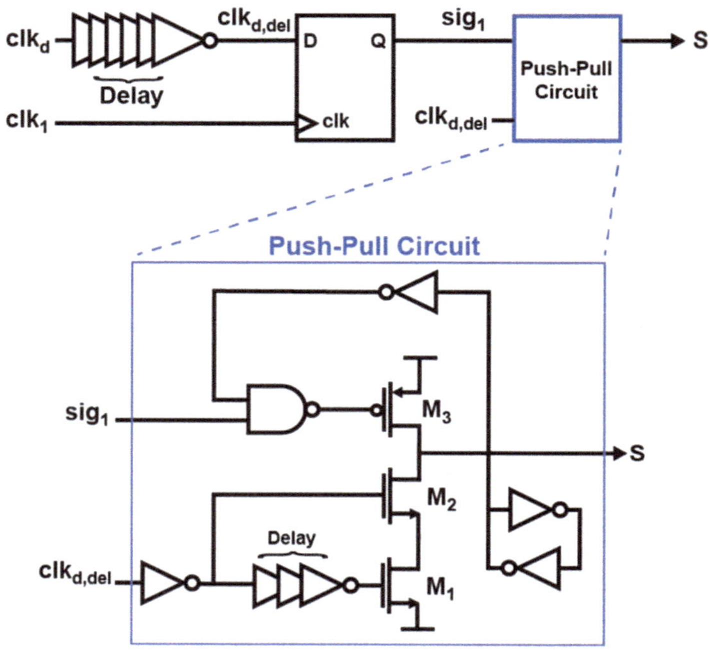

Fig. 5.9 Implementation of the push-pull circuit to generate the masking signal S

that the dead time issue does not have any impact on the output voltage since during the transition of A and $\bar{A}$, the PMOS is masked by the S for the polarity evaluation. Thus, C_{store} is isolated from the switch matrix.

Figure 5.8 shows the implementation of the ring oscillator and the clock divider circuit. clk_1 is used for switching M_1 and M_2 in the inductor converter. In addition, the divider circuit (using D flip-flops) is necessary to set how frequent the TEG polarity is evaluated.

The push-pull stage that generates the masking signal S is shown in Fig. 5.9. As illustrated in the figure, when $clk_{d,\text{del}}$ is high, then sig_1 is low and S is high. In contrast, when $clk_{d,\text{del}}$ goes low, then S is discharged through M_1 and M_2. Hence, the state of S is held by the output latch. This state stays the same till sig_1 is changed. Once sig_1 goes high, then S is pulled high through M_3. Note that M_1 and M_2 can be both ON at the same time due to the delay introduced at M_1 gate.

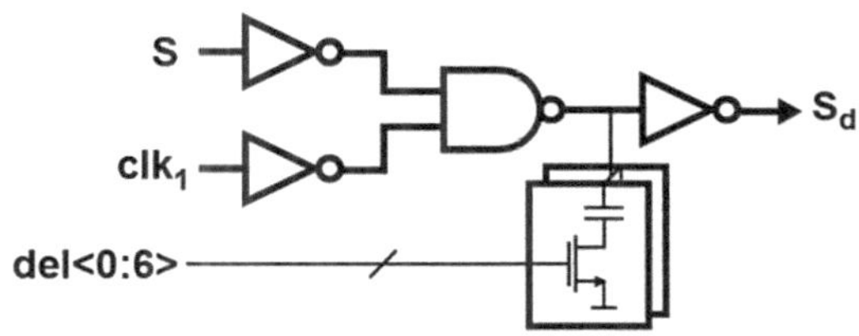

Fig. 5.10 Implementation of the S_d generation circuit

Figure 5.10 shows the delay circuit that generates a delayed version of S which is S_d. S_d triggers the comparison stage to evaluate the state of V_d. S is used to mask the PMOS switch which at the same time clk_1 becomes low and V_d voltage is generated based on the polarity state. Then S_d is generated after some delay to initiate the comparison which allows V_d voltage to settle for reliable evaluation. The timing of S_d is crucial to the operation of the proposed polarity method. Hence, S_d should have enough delay, so that the polarity is evaluated after V_d settles to the value that indicates positive or negative polarity. Finally, extra bits, del(0:6), from a serial register are added to overcome any variations in the circuit that might affect the delay of S_d.

Figure 5.11a displays the comparison methods that are designed to evaluate the state of V_d. As depicted in the figure, two methods are implemented for V_d evaluation where one of them is selected using a multiplexer. A D flip-flop is used to detect the voltage level of V_d. Further, a clocked comparator, which compares V_d with the ground or V_{store}, is utilized for polarity monitoring [90, 91]. Once the comparison is finished, then the state of the output register is updated and saved for the next evaluation. Figure 5.11b depicts the state diagram of the polarity controller. If the polarity is positive, then *sout* is high and the output register holds the previous state. In contrast, if the polarity is reversed, then *sout* changes from high to low which triggers the output register to update the state. The output C will be transmitted to a level shifter and then to the switch matrix to flip the TEG polarity. The two methods of comparison, i.e., D flip-flop and the clocked comparator, are implemented to compare the performance of the two methods which gave the same results for both. However, the comparator has been selected for V_d evaluation in the following sections. Note that during the non-switching cycle of the converter (NMOS and PMOS are off), V_d equals to the inductor voltage in addition to V_{in}. Since the inductor voltage is zero, then V_d equals to V_{in}. Hence, V_d has always a defined state (logic zero) during the non-switching cycles and will not be floating.

Figure 5.12 displays the circuit implementation of the voltage level shifter. The level shifter is used to drive the switch matrix with higher voltage so that their conduction losses and leakage are reduced. The circuit depends on cross-coupled transistors, M_5 and M_6 to pull up the output. Unlike the conventional level shifter, the transistors M_3 and M_4 prevent the contention between the PMOS and NMOS transistors which might occur during the transition.

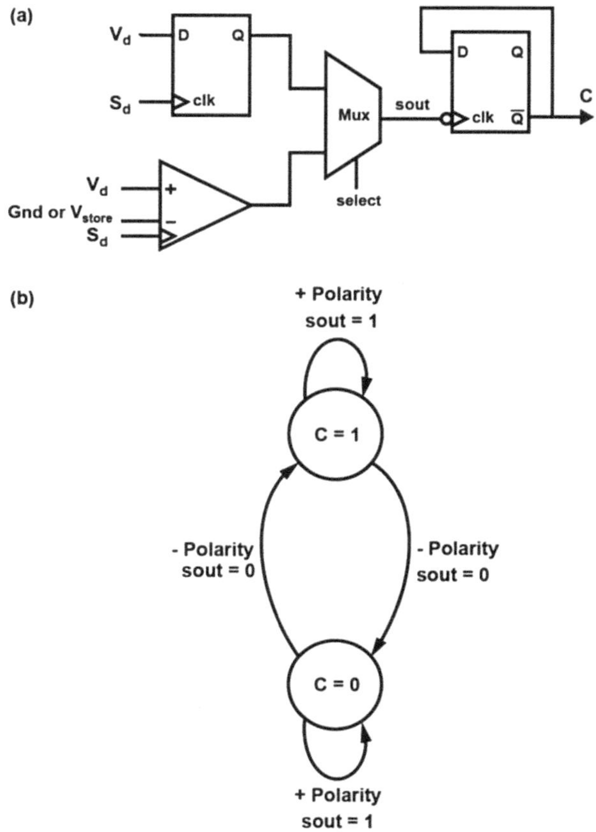

Fig. 5.11 (**a**) Implementation of the comparison stage that evaluates the polarity of the TEG voltage using either a D flip-flop or a clocked comparator. (**b**) State diagram of the polarity controller

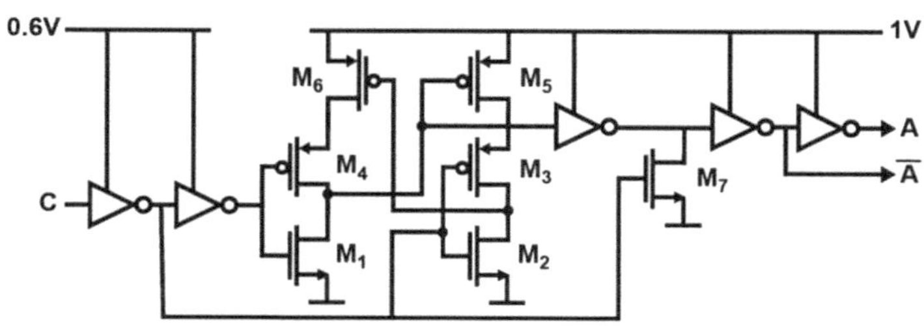

Fig. 5.12 Implementation of the level shifter circuit

5.2.1 Measurement Results of Auto-Polarity Circuits

Figure 5.13 displays a die photograph in 65 nm CMOS and the layout of the thermal harvesting system including the ZCS and the proposed auto-polarity circuit. As depicted in Fig. 5.13a, the overall die area is 9 mm^2 (i.e., 3 mm × 3 mm), and

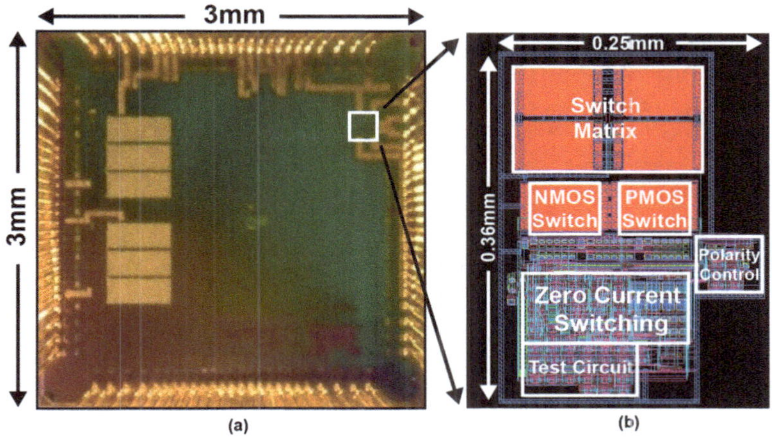

Fig. 5.13 (**a**) Die photograph of thermal harvesting prototype IC in 65 nm CMOS. (**b**) A layout showing different parts of the prototype chip

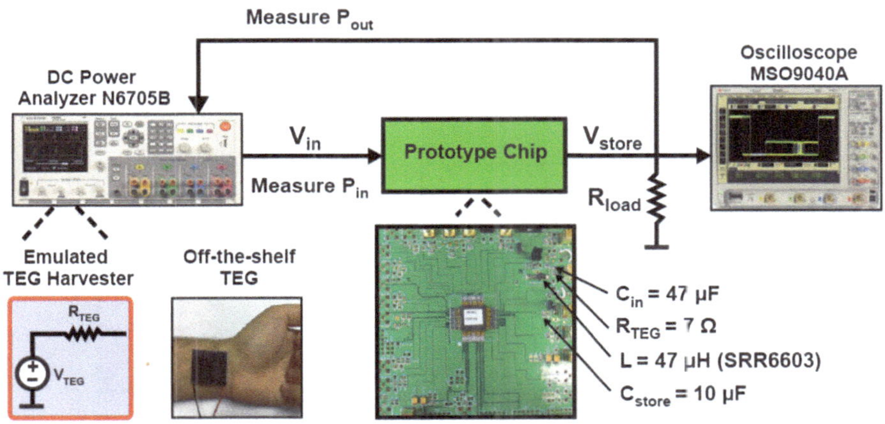

Fig. 5.14 Physical test setup for characterizing the prototype chip

the active area is 0.09 mm^2 (i.e., 0.36 mm × 0.25 mm). The layout in Fig. 5.13b shows different parts of the energy harvesting system. Note that the switch matrix dominates the area while the polarity controller occupies a very small area. Finally, the off-chip components in the system are the inductor, input and output capacitors.

Figure 5.14 shows the test setup used to characterize the performance of the prototype chip. The TEG model is utilized to emulate an off-the-shelf TEG in [34]. A DC power analyzer is used to supply the TEG voltage (20–50 mV) [92]. An external resistor of 7 Ω is utilized to emulate the R_{TEG} [93]. The DC power analyzer is also used to measure the overall efficiency. In addition, the setup includes an oscilloscope to measure the time domain signals. Finally, an off-the-shelf TEG is placed on the author's wrist and connected to the prototype chip to test the performance in a real case scenario.

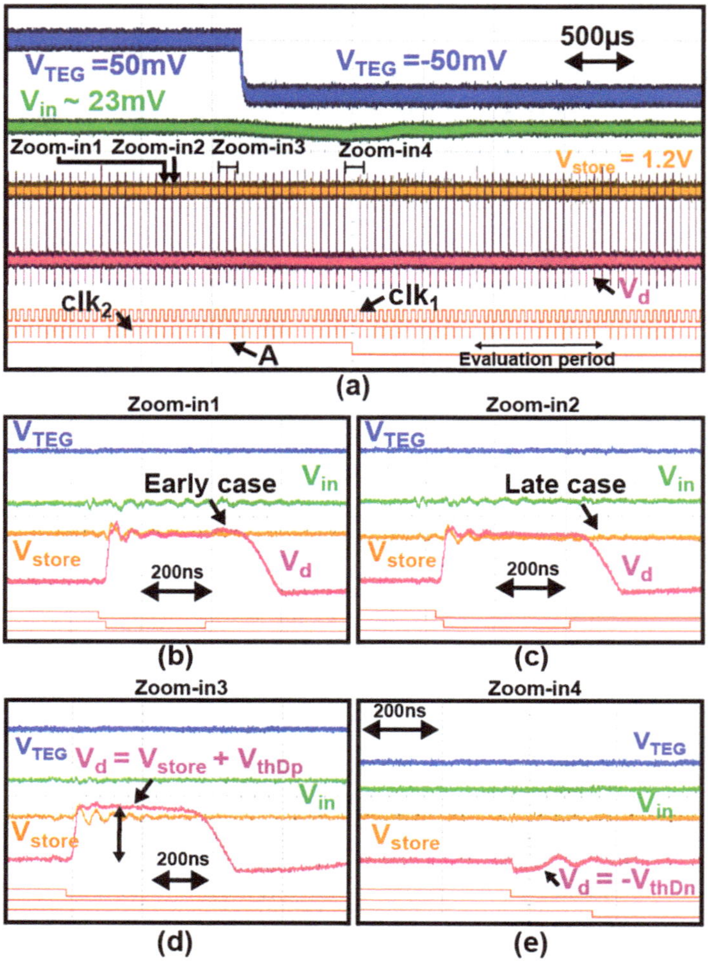

Fig. 5.15 (**a**) Measured time domain signals of the proposed auto-polarity circuit, (**b**) evaluation of ZCS during early condition, (**c**) evaluation of ZCS during late condition, (**d**) evaluation of the input polarity at the positive case, and (**e**) evaluation of the input polarity at the negative case

In the following measured results, first the prototype chip is evaluated when the TEG voltage polarity is changed abruptly, from 50 to −50 mV (in Figs. 5.15 and 5.16). This case might not be realistic since the temperature change is usually slow across the TEG. However, it will help in giving more insight on the dynamics of the system for fast input polarity change. Second, the prototype chip is evaluated when the TEG voltage polarity changes gradually, which provides a more realistic scenario for thermal harvesting (in Figs. 5.17 and 5.18). Finally, an off-the-shelf TEG is placed on the author's wrist and connected to the prototype chip as will be shown in Fig. 5.19 [34].

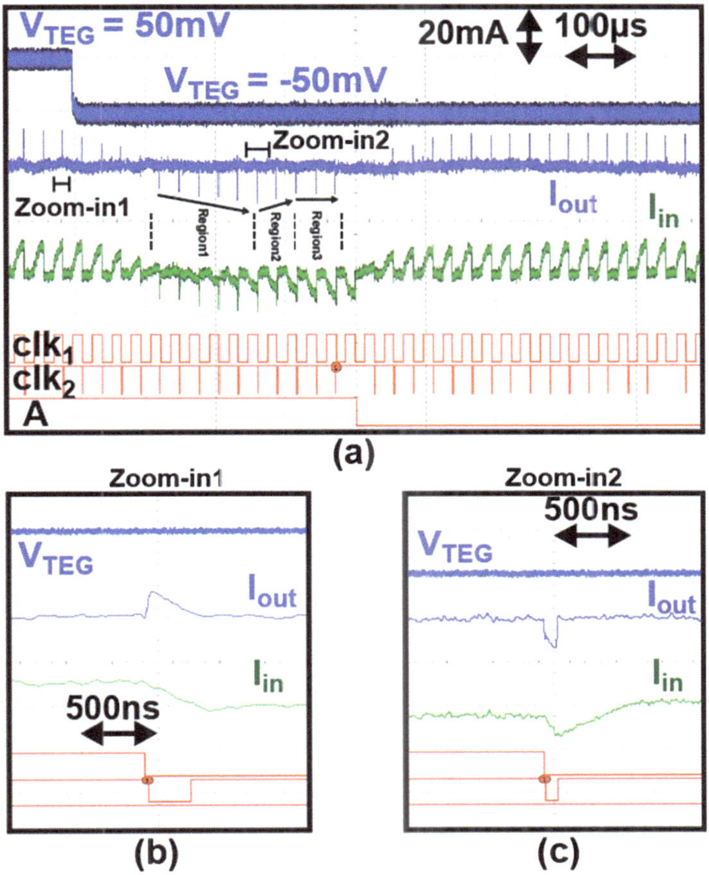

Fig. 5.16 (**a**) Measured time domain signals of the inductor current and the output current of proposed auto-polarity circuit, (**b**) zoom-in version of the current during normal operation, and (**c**) zoom-in version of the current when the polarity is flipped

Figure 5.15a shows the time domain signals of the inductor converter when the TEG voltage is changed abruptly from 50 to −50 mV. Initially, V_{TEG} voltage is 50 mV and V_{in} is approximately half V_{TEG}. The polarity is evaluated every 16 cycles of clk_1, where clk_2 is masked to prevent discharging C_{store}. Figure 5.15b, c shows the operation of the ZCS circuit. As depicted in the figures, clk_2 toggles around the zero current point based on the state of the V_d voltage, either early or late case. In addition, V_{TEG} is changed from 50 to −50 mV in between two evaluation cycles. Hence, the polarity controller will detect and reverse the input voltage in the next evaluation cycle. Note that A is triggered at the evaluation cycle to correct for the reversed polarity using the switch matrix as indicated in Fig. 5.15a. Finally, V_{TEG} graph in Fig. 5.15a is shifted up for better clarity.

Figure 5.15d, e shows the state of V_d voltage at the positive and the negative TEG voltage, respectively. As indicated in Fig. 5.15d, which is the zoom-in3 region

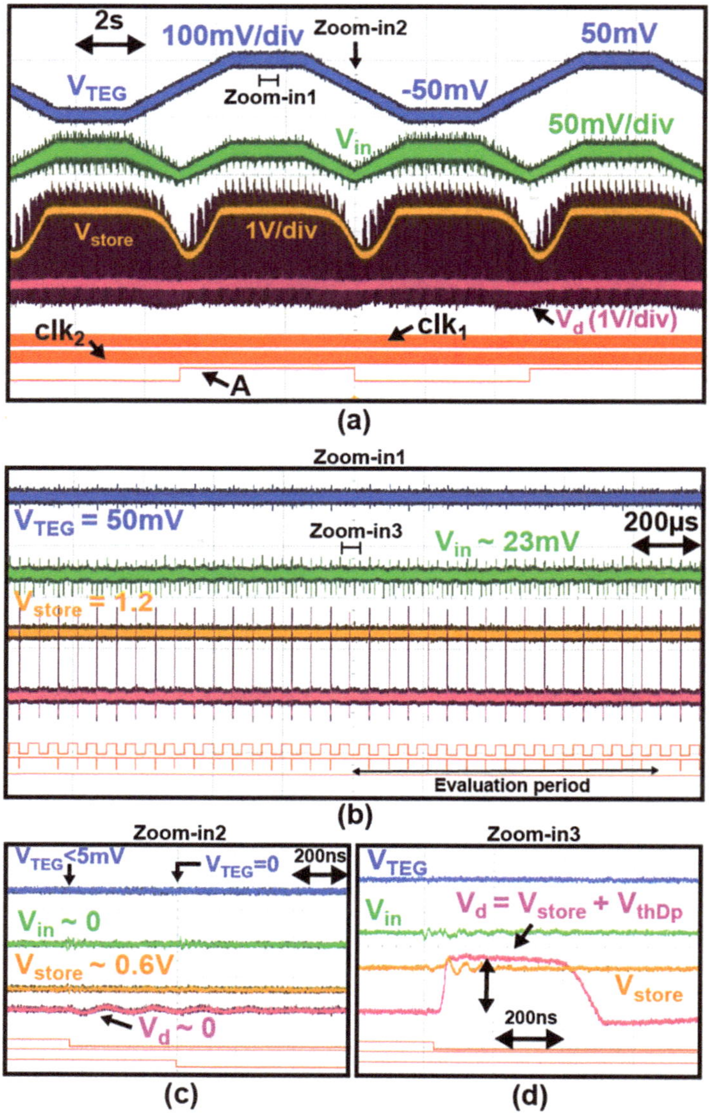

Fig. 5.17 (**a**) Measured time domain signals of the proposed auto-polarity circuit, (**b**) zoom-in version of the circuit during normal polarity, (**c**) zoom-in version when the input voltage near to zero value, and (**d**) evaluation of the input polarity at the positive case

in Fig. 5.15a, the V_d is pumped to $V_{\text{store}} + V_{THD_p}$ which indicates a positive input voltage. In contrast, Fig. 5.15e, which is the zoom-in4 region in Fig. 5.15a, shows the V_d voltage is equal to $-V_{THD_n}$ ($\sim -300\,\text{mV}$) which indicates a negative input voltage.

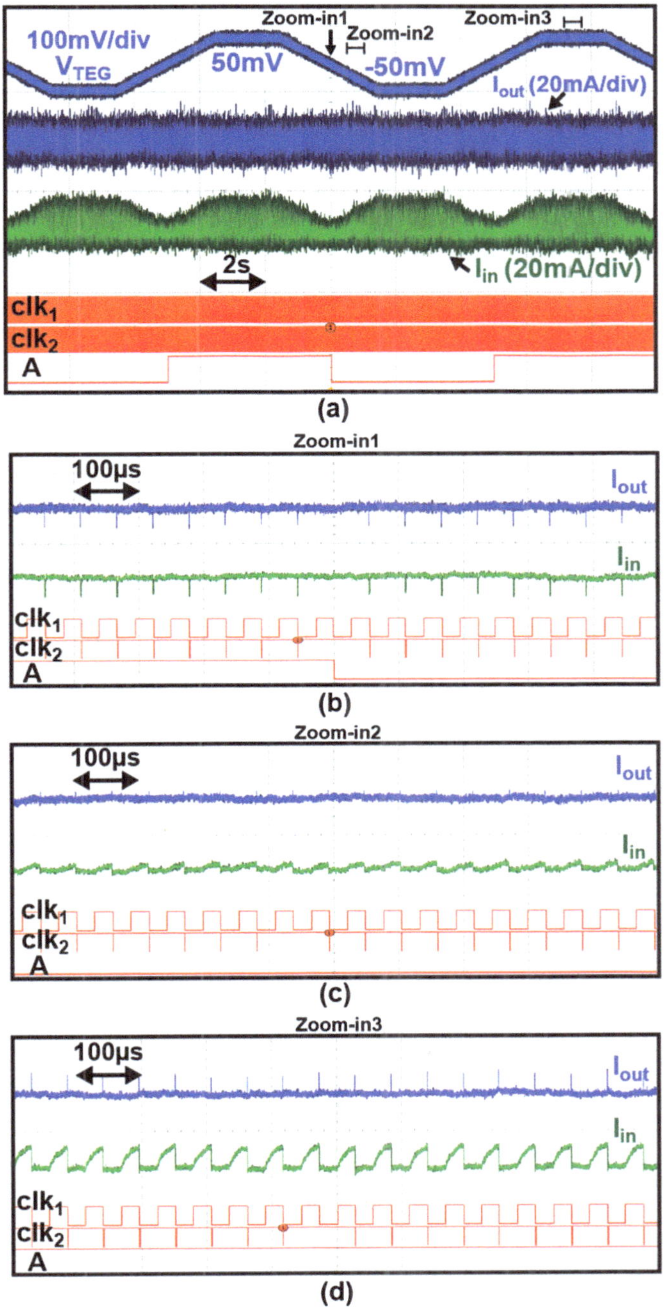

Fig. 5.18 (**a**) Measured time domain signals of the inductor and the output current of the proposed auto-polarity circuit, (**b**) zoom in of the currents when the TEG voltage near to zero, (**c**) zoom-in version when the TEG voltage starts to charge the inductor, and (**d**) zoom-in version of the currents during normal operation

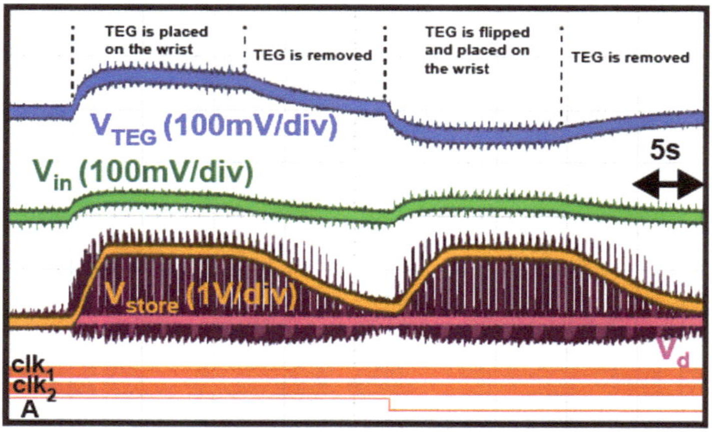

Fig. 5.19 Measured time domain signals of the proposed circuit using an off-the-shelf TEG placed on the author's wrist as was shown in Fig. 5.14

Figure 5.16 shows the measured inductor current and the output current through C_{store} when V_{TEG} goes from the normal to the reversed polarity case. The current measurement is performed by placing two resistors (1 Ω each), one before the inductor and the other before C_{store}. During the normal polarity, the boost converter works normally where the inductor is charged during clk_1 on-time and then discharged into C_{store} during clk_2 on-time as shown in Fig. 5.16b. Once the polarity is flipped, then the inductor current reverses its direction and is charged by the input voltage during the on-time of clk_1. When clk_1 becomes low, then clk_2 becomes low as well and the inductor is charged from C_{store}. Thus, a large jump in the inductor current is produced as indicated in Fig. 5.16c. In region 1 shown in Fig. 5.16a, $|I_{out}|$ is increasing since the inductor current is rising as was also explained in Fig. 5.6d. When $|I_{out}|$ reaches the maximum value, then it will start to decrease as indicated in region 2 since ZCS senses negative current. Hence, the ZCS will reduce the pulse width of clk_2 which will help in decreasing the negative current going back to the boost converter. However, in the case shown in Fig. 5.16a, the ZCS saturates in region 3 to the minimum clk_2 pulse width [94]. Thus, $|I_{out}|$ stays the same amplitude in region 3. Note that the ZCS works in the favor of the polarity method since it reduces the energy depleted from C_{store} during the polarity flip.

Figure 5.17a shows the time domain signals of the inductor converter when V_{TEG} is changed gradually from 50 to −50 mV. V_{TEG} voltage is changed from positive to negative in more than 2 s. Hence, V_{store} is discharged during this transition. Figure 5.17b, which is the zoom-in1 in Fig. 5.17a, shows the normal operation of the inductor converter during the positive polarity case. Figure 5.17c, which is the zoom-in2 in Fig. 5.17a, displays the time domain signals when polarity controller corrects the polarity of the TEG. At this particular case, A signal is low which indicates a negative polarity case. This means that if the V_{TEG} voltage is less than 5 mV, then the inductor current will be very small which results in almost zero V_d

voltage. Thus, the polarity controller will consider it as negative as was shown in Fig. 5.11. As such, the proposed auto-polarity system can detect the polarity of the V_{TEG} if it is higher than 5 mV. Finally, Fig. 5.17d, which is the zoom-in3 in Fig. 5.17b, shows the case during the evaluation of the positive polarity.

Figure 5.18a shows the time domain signals of the inductor and output current when V_{TEG} polarity is changed gradually from 50 to −50 mV. In Fig. 5.18b, which is the zoom-in1 in Fig. 5.18a, V_{TEG} is almost zero and C_{store} is discharging during clk_2 pulse. Further, as explained in Fig. 5.17c, A is flipped when V_{TEG} voltage is less than 5 mV. Figure 5.18c shows the inductor current when it starts to build up and transfer to C_{store}. Finally, Fig. 5.18d shows the inductor current and the output current during normal operation.

Figure 5.19 shows the time domain signals of the boost converter when an off-the-shelf TEG device is connected to the prototype chip [34]. This measurement is performed at room temperature and the TEG is placed on the author's wrist. As depicted in the figure, the positive TEG side is placed on the wrist and the system generates an output voltage of 1.2 V. Then the TEG is removed for 10 s. During this time, the output voltage is decreasing since C_{store} is depleting. After that, the TEG side is flipped and placed again on the wrist. Hence, the polarity controller corrects the polarity and the system is able to harvest from the TEG. Note that if the output voltage becomes zero during the polarity change, then the system needs to startup again. For the sake of testing the polarity technique, we assume a voltage already available for the control logic of the TEG.

Figure 5.20 displays the measured efficiency and the output voltage at different output loads and V_{TEG} voltages. As depicted in Fig. 5.20a, the maximum efficiency

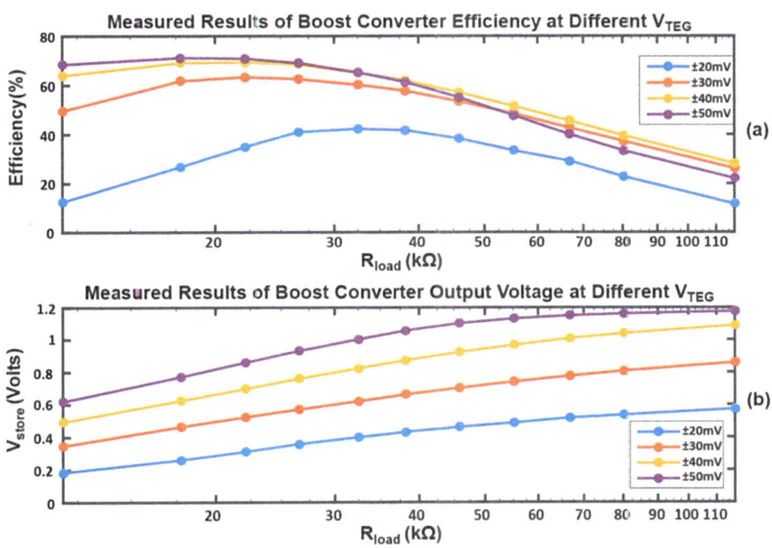

Fig. 5.20 (**a**) Measured results of the end-to-end efficiency of the overall system. (**b**) Measured results of the inductor converter output voltage

Table 5.2 Comparison with other published work

Parameter	JSSC'15 [10]	PowerMEMs '09 [13]	Linear Tech. '12 [20]	ISELPED'14 [30]	ISELPED'15 [31]	Work in this book
Technology	130-nm CMOS	Not applicable	Not reported	350-nm CMOS	Not applicable	65-nm CMOS
Min. voltage for polarity	Not applicable	±200 mV	±30 mV	±40mV	Not applicable	±5 mV
Min. working voltage	+10 mV	±200 mV	±30 mV	±40 mV	+16mV	±20 mV
Maximum output voltage	1.1 V	2.5 V	5 V	5.5 V	3 V	1.2 V
Peak efficiency	83% at 300 mV	70% at 200 mV	55% at 70 mV	61% at 70 mV	53% at 50 mV	70% at 50 mV
Output power	22 mW	3.85 mW	25 mW	470 μW	1.7 mW	40 μW
Auto-polarity method	Not implemented	Direct input comparison	Two parallel transformers	Transformer	Not implemented	Evaluation of V_d voltage
Circuit implementation	On-chip	Off-chip (PCB)	On-chip	SPICE simulation	Off-chip (PCB)	On-chip
Startup mechanism	RF	Not implemented	Positive feedback	Positive feedback	Piezoelectric transformer	External supplies

is 70% at V_{TEG} of 50 mV, output current of 40 μA, and output voltage of 0.9 V. Note that the power consumption from the 0.6 V supply is 200 nW. Further, the power consumption from the 1 V supply is 1 μW. The power consumption from both the supplies is included in the measured efficiency. Finally, Fig. 5.20 is also valid for negative value of V_{TEG} voltages.

Table 5.2 shows the performance of the proposed auto-polarity compared to recent work. As shown by the table, the prototype chip presented in this paper is the only auto-polarity method that can be integrated on-chip compared to the direct input comparison method reported in [86] and transformers technique reported in [87] and [95]. The proposed technique achieves the minimum voltage for polarity detection (∼5 mV) compared to 30 mV, 40 mV, and 200 mV in [87], [95] and [86], respectively. In [87] and [95], the minimum voltage depends on the transformer ratio used. However, in the proposed polarity, the minimum voltage depends on the amplitude value of V_d that can be detected using the polarity controller. The working voltage range in our system is between 20 and 50 mV due to the available power from the TEG needed for the load.

The peak efficiency of the proposed system is 70% at 50 mV TEG voltage. In [86] and [96], which both have off-chip implementation, they achieve 70% at 200 mV and 53% at 50 mV, respectively. For transformer-based design in [87] and [95], they achieve 55% at 70 mV and 61% (SPICE simulation) at 70 mV, respectively. In particular case, [87] has lower efficiency than the proposed system since it presents a complete chip that has more complex control circuits with more components in addition to voltage references and regulators at different CMOS technology. Hence, more circuit elements inclusion in their chip will increase wasted energy and reduce efficiency. Finally, [97] has higher efficiency since the design uses current control scheme and duty cycled with offset compensated comparator which helped in maintaining high efficiency.

Chapter 6
Energy Combiner and Power Manager for Multi-Source Energy Harvesting

6.1 Reported Techniques in Energy Combiner Techniques

There are many energy combining techniques that are reported in the literature for multi-source energy harvesting systems which are summarized in Fig. 6.1. As depicted in Fig. 6.1a, a time multiplex scheme is applied to multiple energy sources [78]. As such, different time periods are given to each energy source to transfer its energy to the output load. The system in Fig. 6.1a shares the power conversion stage with all energy sources which reduces the cost of the system. Further, time multiplexing provides the flexibility to implement MPPT circuit for each energy source within the given time slot, however, it assumes that the energy from all sources is available at all times (which is not the case in many applications). Having said that, if one source is not available, the stored energy in the storage capacitor could be depleted back to the source which might cause the output voltage and efficiency to be degraded.

Another technique for energy combiner is shown in Fig. 6.1b. In this design, a switched capacitor (SC) charge pump is utilized to stack storage capacitors in series so that their individual voltages are summed [98]. When voltages from different harvesters are summed, then the total voltage should not exceed the breakdown voltage of the transistor. Further, care should be taken when one source is not available in such implementation. Finally, it is hard to maintain a high efficiency for the SC charge pump over a wide range of input voltage and output load.

Figure 6.1c shows a voltage level detection method to combine energy from different sources. It simply compares the voltage level from each harvester and chooses the one with the higher voltage level [71, 99–101]. However, higher voltage does not necessarily mean higher power which results in not utilizing the energy sources efficiently.

Figure 6.1d displays the power oring technique to combine energy from multi-energy source. It uses simply a diode from each source to transfer the energy to the output storage [102–104]. This method is the simplest form of energy

M. Alhawari et al., *Energy Harvesting for Self-Powered Wearable Devices*,
Analog Circuits and Signal Processing, DOI 10.1007/978-3-319-62578-2_6

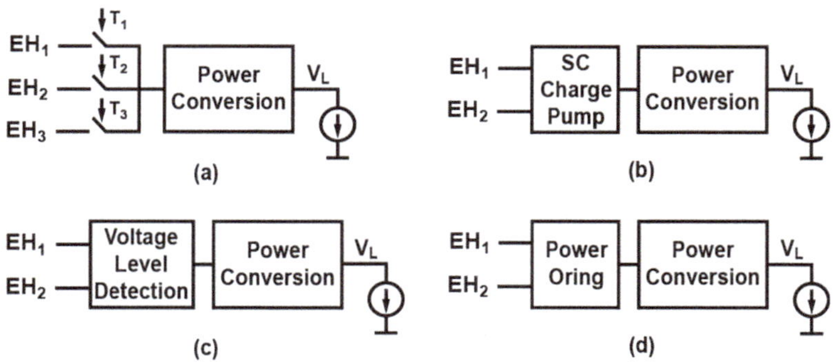

Fig. 6.1 Recent work of energy combiner techniques. (**a**) Time multiplex scheme [78]. (**b**) Using SC charge pump [98]. (**c**) Detecting the highest voltage level [71, 99–101]. (**d**) Power oring using diodes [102–104]

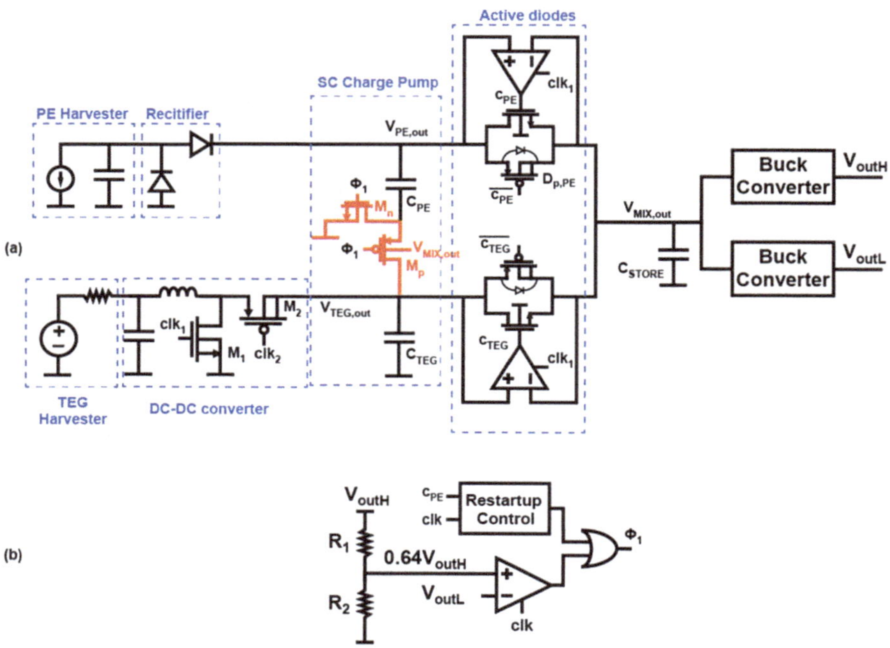

Fig. 6.2 Proposed power oring combiner with re-startup prevention

combiner. However, the drop voltage of the diodes might degrade the efficiency of the combiner. Instead, active diodes could be used to enhance the efficiency.

A multi-source energy harvesting combiner is shown in Fig. 6.2 as part of a complete energy harvesting system. The proposed combiner addresses the limitations of the present art as follows.

- The system integrates the ability of startup mechanism from one of the passive harvesters without adding extra components and by using the parasitic diode of the PMOS switch in the energy combiner. As such, decreasing complexity and area overhead.
- Active diodes are used to transfer the harvested energy to the storage capacitor. This helps in eliminating the voltage drop from using the Schottky diodes. Hence, higher efficiency is achieved and larger energy is available at the output storage.
- Two switches are used for re-startup prevention by summing the harvesters voltages using their output capacitors based on monitoring the regulated voltages from the buck converters. As such, maximizing the power generated from the harvesters and preventing the re-startup cycle and using the harvesters even if their individual voltage is not enough to run the load.

As indicated in Fig. 6.2, there are two energy harvesting sources; namely, TEG and piezoelectric (PE) harvester. Two buck converters are connected to the energy combiner to generate two regulated voltages, V_{outH} and V_{outL}, where V_{outH} is larger than V_{outL}. Having two regulated voltages is common in wireless sensor nodes as the V_{outH} can be used to run SRAM, transmitter, sensors, etc., and V_{outL} is used to operate the core of the device to reduce the power consumption.

Figure 6.2a shows a detailed circuit of the energy combiner as part of a complete energy harvesting system. As depicted in the figure, each energy source is connected to an active diode. An active diode behaves like a diode but without the issue of the diode voltage drop. It consists of a transmission gate (NMOS and PMOS switches) and a clocked comparator. The clocked comparator is used to eliminate the static current and reduce the power consumption compared to the continuous-time comparator. The clock of the comparator is chosen to be the clock of the DC–DC converter such that the evaluation of the clocked comparator is done during the charging phase of the DC–DC converter. The PMOS switch in the transmission gate is connected such that its parasitic diode $D_{p,PE}$ is connected from the input to the output, which is essential for startup mechanism using one of the energy sources. A re-startup prevention technique is used to prevent the re-startup cycle of the system by connecting both capacitors in series based on a control signal Φ_1. Φ_1 signal is generated only when the voltage of the individual source is not enough for the output load. Figure 6.2b shows Φ_1 generator circuit, it consists of a voltage comparator that monitors the regulated output voltage V_{outL} and a divided version of V_{outL}. In addition, Fig. 6.2b shows the re-startup control logic that generates Φ_1 based on a predetermined frequency.

The basic operation of the energy combiner is working as the following. Initially, all energy storages have zero charges. The startup mechanism is done from one of the energy sources that use passive converter which is a PE harvester in this case. This will result in charging up C_{PE} storage. Further, due to the PMOS connection in the active diode and its parasitic diode $D_{p,PE}$, charges will be transferred to the main storage element C_{STORE}. After some time, C_{STORE} will accumulate charges so that enough voltage is generated for the buck converters to generate V_{outH} and V_{outL}, where V_{outH} is larger than V_{outL}. These two voltages are used to drive the load as well as the internal clock needed to power the DC–DC converter, the clocked comparators, and the control circuits.

During normal operation, the active diodes pass the voltages generated from the energy sources as long as these voltages are higher than $V_{MIX,out}$. As such, $V_{MIX,out}$ will be high enough to generate both V_{outH} and V_{outL}. Therefore, the Φ_1 signal will be logic 1 so that both capacitors C_{PE} and C_{TEG} are separated by the switches. In case the $V_{MIX,out}$ becomes lower than V_{outH} but still higher than V_{outL}, then the clocked comparator in Fig. 6.2b is triggered. This indicates that the voltage generated from the individual energy source is not enough to generate the regulated voltage V_{outH}. As such, the signal Φ_1 starts to alternate between 0 and 1. During 0 logic, the two capacitors C_{PE} and C_{TEG} are in series and both voltages are summed at $V_{PE,out}$ node and becomes $V_{PE,out} + V_{TEG,out}$. Hence, the top active diode is activated to transfer the charge to the storage capacitor C_{STORE}. Once the charges are transferred, the top active diode is turned off and the re-startup control forces Φ_1 again to logic 1 so that the capacitors C_{PE} and C_{TEG} are separated again to be charged by their harvesters to their corresponding voltages. Then, Φ_1 keeps alternating between 1 and 0 till $V_{MIX,out}$ is high enough to generate V_{outH} again. Note that the system uses the charge pump for a certain period of time so that the output voltage is recovered, unlike [98], the system ensures that the voltage generated by the charge pump does not exceed the voltage rating of the transistor using the control logic shown in Fig. 6.2b.

6.2 Power Manager Implementation for Multi-Source Energy Harvesting

In this section, a multi-source energy harvesting architecture that supports low power devices will be presented. Figure 6.3 displays the complete system design of the energy harvesting system. It includes the energy combiner explained in the last section which integrates two energy harvesters and two buck converters. The inductor-based boost converter contains the ZCS technique and auto-polarity method explained in Chaps. 4 and 5, respectively. The buck converters are assumed to generate two regulated voltages, namely 0.6 and 1 V [61]. A power manager (PM)

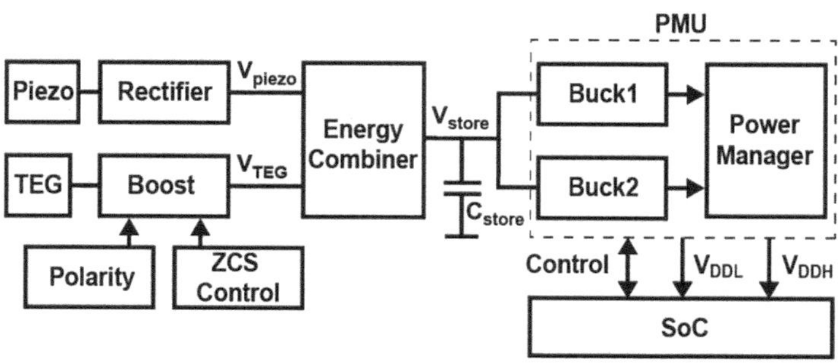

Fig. 6.3 Block diagram of the energy harvesting system including the PMU

manages the power delivered to the load and generates control signals to produce different modes of operation based on the available power from the harvesters. The two buck converters and the PM form the power management unit (PMU) of the system. The load is assumed to be a system-on-chip (SoC) that includes an analog front end (AFE), analog to digital converter (ADC), and an ECG processor. The SoC samples ECG signals and processes the ECG data to extract different features that indicate abnormality in the heart [85, 105]. The AFE/ADC operates at 0.6 V. In contrast, the ECG processor uses the 0.6 V for the digital core, while the 1 V is to operate the SRAM. Note that the PM should work at the lowest possible voltage in the system, i.e., 0.6 V.

6.2.1 Biomedical Processor

The biomedical processor is an ECG processing system that performs ECG feature extraction to indicate an abnormality in the heart [85]. As indicated in Fig. 6.4, the processor consists of a processing core and an SRAM. Multiple voltage designs are used to minimize the overall power consumption of the system. Voltage islands are created for the SRAM and the processing core as shown in Fig. 6.4. As such, having separate supply voltages for the SRAM and the processing core would reduce the overall power dissipation of the system. Note that buffers, level shifters, and isolation cells are needed due for implementing multiple voltage domains in any system.

Clock gating enables power saving by reducing the dynamic power. Since all the operations in the processor core are not performed at the same time, then some blocks could be clock gated while other blocks are running. Figure 6.4 shows the clock gating setup where it is controlled by the power manager.

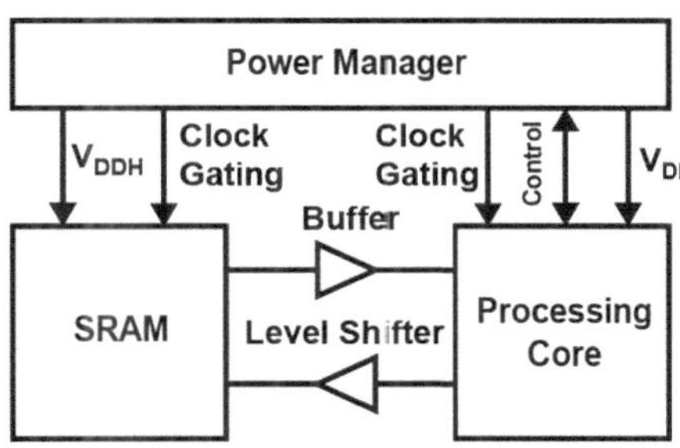

Fig. 6.4 Power manager design to support power and clock gating

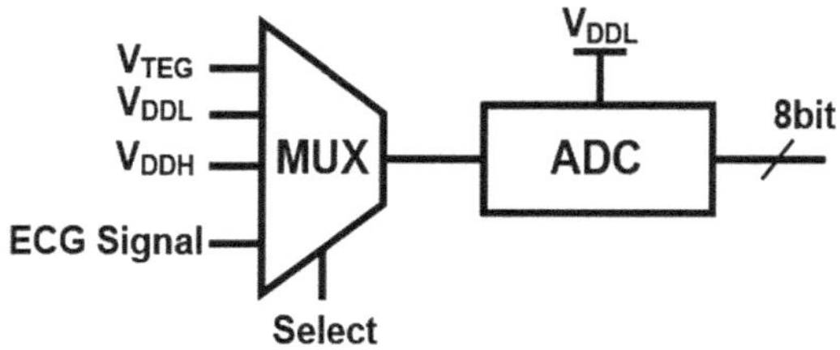

Fig. 6.5 Voltage monitoring by using ADC and selecting the input using a Mux

6.2.2 Power Manager

The power manager (PM) module is an *always-on* digital circuit that controls the power flow from the energy source to the load. The PM ensures a proper operation of the overall system considering the available stored energy. Since the PM is an always-on circuit, it should be powered from the minimum available output voltage which is in our case the 0.6 V (V_{DDL}). Different control signals can be generated from the PM for different purposes to the processor such as the power gating and clock gating signals. The main component in the PM is the voltage supervisory circuit that monitors the energy available from the harvesters. Figure 6.5 shows the voltage monitoring scheme. As indicated in the figure, an ADC can be used to digitize three voltages from the energy harvesters, namely V_{TEG}, V_{DDL}, and V_{DDH} which represent the boost converter output voltage, the regulated 0.6 V and 1 V, respectively. In addition, the ECG signal is digitized using the same ADC and by using a multiplexer. Note that the ADC is an always-on block which is powered from the same supply of the PM, i.e., V_{DDL}. Finally, the rate of which the voltage is monitored for the PM should not affect the ECG sampling rate.

To gain an understanding of the power domains in the SoC, Fig. 6.6 shows the power domains needed for different blocks in the SoC. As indicated in the figure, there are two blocks, always-on blocks and switchable blocks. The always-on blocks include the PM and the AFE/ADC which are powered by the lowest voltage from the PMU, i.e., 0.6 V. In contrast, the switchable blocks are divided into two parts, high energy mode blocks and low energy mode blocks. The high energy mode blocks are the SRAM and a digital core. This mode is activated when there is enough energy from the harvesters to generate the two regulated voltages. Therefore, the SoC runs at the maximum performance and activates all the functionalities. In low energy mode blocks, the SoC activates only a low power accelerator that is powered by the 0.6 V supply. This mode indicates that the harvesters are only able to generate the 0.6 V supply which indicates a low power mode.

Figure 6.7 shows the state diagram of the FSM implemented for the PM. As indicated in the figure, there are six states for the PM. It starts by the first mode M_1 where the system waits for startup from the passive harvester which is the PE in this particular case. Once the voltage level is adequate at the storage capacitor C_{store}, then the switched capacitor will be turned on to provide the first regulated voltage V_{DDL}. Hence, in M_2, the PM, ADC, and the boost control will be turned on to initiate

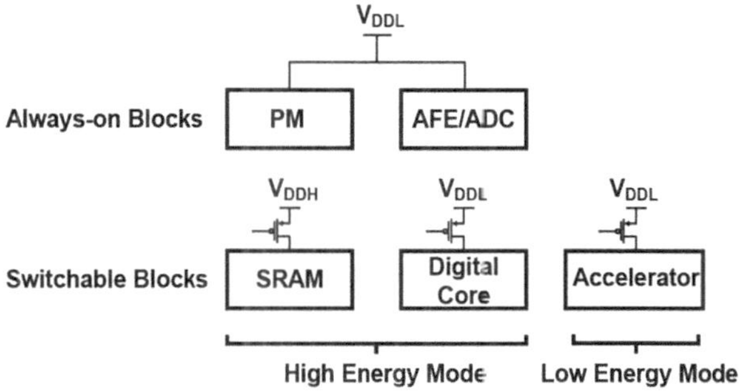

Fig. 6.6 Power domains for different parts of the SoC which includes always-on blocks and switchable blocks

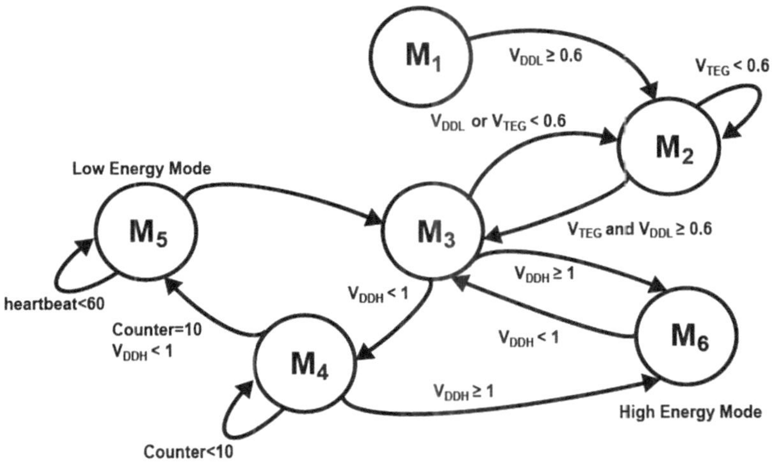

Fig. 6.7 State diagram of the finite state machine for the PM

the control signals for different parts of the SoC. Further, the PM will monitor the voltage from the boost to make sure that it is sufficient enough for harvesting since the TEG is the main source of the energy in the overall system. Therefore, if V_{TEG} is less than 0.6 V, then the PM will send a signal to the polarity control to check the input polarity and reversed if needed. Once the V_{TEG} is higher than 0.6 V, then the PM will move to M_3 for continuous monitoring.

During M_3 state, V_{DDH} is monitored which decides if the SoC will work in high or low energy mode. If V_{DDH} is larger or equal to 1 V, then the SoC enters the high energy mode. Hence, the SoC will extract ECG feature and run complete functionalities. In contrast, if V_{DDH} is less than 1 V, then the PM will enable the switches in the energy combiner to sum both the harvesters voltages. If the summation is able to increase V_{DDH} to 1 V, then the SoC will go to high energy mode again, otherwise, the SoC enters the low energy mode M_5. In M_5, a low power accelerator is activated to measure the heart beats. In addition, after some time, the PM will check again the level of V_{DDH} to decide the energy mode again [106].

Table 6.1 Summary of mode of operation for the PM

Modes	V_{TEG}	V_{0_6}	V_1	PM/ACD	SRAM/core	Accelerator	Notes
M_1	x	<0.6	x	Off	Off	Off	Waiting for startup
M_2	<0.6	≥0.6	x	On	Off	Off	SoC off, Boost on. Polarity check
M_3	≥0.6	≥0.6	x	On	Off	Off	Monitoring
M_4	x	x	<1	On	Off	Off	Combiner on
M_5	x	x	<1	On	Off	On	Low energy mode
M_6	x	x	>1	On	On	Off	High energy mode

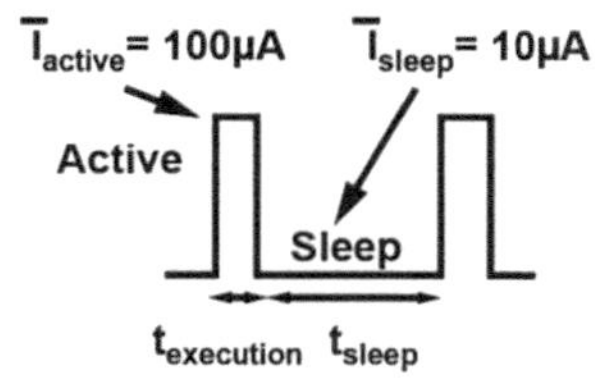

Fig. 6.8 Sleep mode operation of the ECG processor

To summarize the modes of operation, Table 6.1 displays the power state of each block at each mode. As indicated in the table, the PM and ADC are always on and are operated at the lowest regulated voltage. Further, M_5 and M_6 represent the low and high energy modes, respectively.

6.2.3 Sleep Mode Operation

The sleep model is a scheme where the processor is made active for a certain period of time to execute its operation, then it goes to sleep to reduce the power consumption. Typically the duty cycle of on to off period is very small, around 2–5%. Using such scheme, a high power load can be run even if the capability of the harvester is lower than needed. This means most of the energy consumed by the load is supported by the storage capacitor.

As an example to explain this scheme, the ECG processor is assumed to have an active and a sleep current consumption of 100 and 10 µA, respectively. Further, the current capability of the harvesters is assumed to be 50 µA. Figure 6.8 shows the sleep mode of the processor. At 1 V, the processor needs 15 ms as execution time $t_{\text{execution}}$, at 1 MHz to execute its algorithm and to analyze three heart beats [107]. Then the processor will wait for 2 s, t_{sleep}, to acquire the next heart beats samples. As such, the processor will go to high energy mode during the active period and then go to sleep for the next 2 s.

The sizing of the storage capacitor C_{store} is essential for setting how much voltage drop is allowed on V_{store}. To explain, let's first assume that we allow 50 mV voltage drop on V_{store} node during the activity of the ECG processor. Thus, the capacitance needed for achieving that voltage drop is shown in Eq. (6.1)

$$C_{\text{store}} \geq \frac{100\,\mu\text{A} \times 15\,\text{ms}}{50\,\text{mV}} \rightarrow C_{\text{store}} \geq 30\,\mu\text{F} \tag{6.1}$$

We can choose C_{store} to be 47 μF with the same package of 30 μF. As such, with 47 μF capacitance, the voltage drop becomes 32 mV as shown in Eq. (6.2)

$$dV \geq \frac{100\,\mu\text{A} \times 15\,\text{ms}}{47\,\mu\text{F}} \rightarrow dV = 32\,\text{mV} \tag{6.2}$$

Now we can calculate how much time is needed to recover the drop in voltage as shown in Eq. (6.3) where 50 μA is the current available from the harvester.

$$t_{\text{recover}} \geq \frac{47\,\mu\text{F} \times 32\,\text{mV}}{50\,\mu\text{A} - 10\,\mu\text{A}} \rightarrow t_{\text{recover}} = 37.6\,\text{ms} \tag{6.3}$$

Given the analysis above, the energy harvester can support a 100 μW for 15 ms execution burst for the ECG processor every 37.6 ms. Comparing the sleep model shown in Fig. 6.8, the sleep time of 2 s is much larger what the capacitor needs to recover the drop in voltage which is 37.6 ms.

The frequency of operation of the ECG processor affects the execution and sleep time. Figure 6.9 shows the relationship between the execution time of the ECG processor and the clock frequency. As indicated in the figure, the minimum operation frequency is 16 kHz which results in a 1.875 s execution time. Note that at this frequency, the sleep time is 0.125 s which is enough for the storage capacitor to recover from the drop of the voltage.

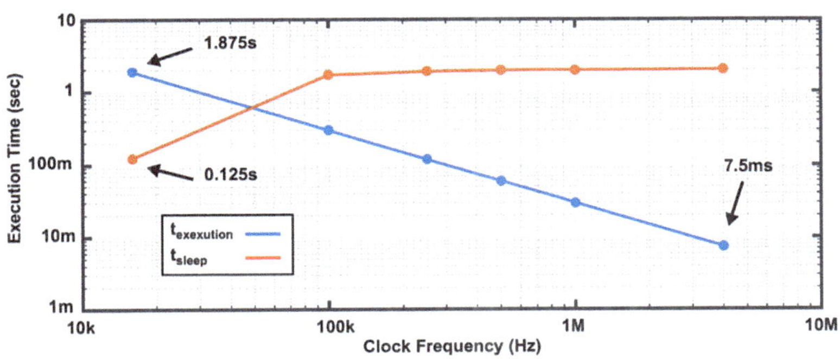

Fig. 6.9 Execution time of the ECG processor for each clock frequency

References

1. G. Girardin et al., Technologies and sensors for the internet of things, October 2014
2. T. Danova, The internet of everything, October 2014
3. PsiKick, The road to trillions (2015)
4. T. Coughlin, The internet of things meets the storage of everything [the art of storage]. Consum. Electron. Mag. IEEE **4**, 118–120 (2015)
5. A. Wang, J. Kwong, A. Chandrakasan, Out of thin air: energy scavenging and the path to ultralow-voltage operation. IEEE Solid-State Circuits Mag. **4**, 38–42 (2012)
6. A. Chandrakasan, N. Verma, C. Denis, Ultralow-power electronics for biomedical applications. Annu. Rev. Biomed. Eng. **10**, 247–274 (2008)
7. J. Emil, M. Aleksandar, O. Chris, C. Piet, A wireless body area network of intelligent motion sensors for computer assisted physical rehabilitation. J. Neuroeng. Rehabil. **2**, 6 (2005)
8. G. Ottman, H. Hofmann, A. Bhatt, G. Lesieutre, Adaptive piezoelectric energy harvesting circuit for wireless remote power supply. IEEE Trans. Power Electron. **17**(5), 669–676 (2002)
9. D. Puccinelli, M. Haenggi, Wireless sensor networks: applications and challenges of ubiquitous sensing. IEEE Circuits Syst. Mag. **5**(3), 19–31 (2005)
10. R. Vullers, R. Schaijk, H. Visser, J. Penders, C. Hoof, Energy harvesting for autonomous wireless sensor networks. IEEE Solid-State Circuits Mag. **2**(2), 29–38 (2010)
11. J. Kwong, A. Chandrakasan, An energy-efficient biomedical signal processing platform, in *2010 Proceedings of the ESSCIRC*, Sept 2010, pp. 526–529
12. N. Verma, A. Shoeb, J.V. Guttag, A.P. Chandrakasan, A micro-power EEG acquisition SoC with integrated seizure detection processor for continuous patient monitoring, in *Symposium VLSI Circuits* (2009), pp. 2–3
13. A. Agnes, E. Bonizzoni, P. Malcovati, F. Maloberti, A 9.4-ENOB 1V 3.8μW 100kS/s SAR ADC with time-domain comparator, in *IEEE Solid-State Circuits Conference Digest of Technical Papers* (2008), pp. 246–248
14. Texas Instruments, CC2550 low-cost low-power 2.4 GHz RF transmitter (2013), http://www.ti.com/lit/ds/swrs039b/swrs039b.pdf
15. S. Bandyopadhyay, P. Mercier, A. Lysaght, K. Stankovic, A. Chandrakasan, A 1.1 nW energy-harvesting system with 544 pW quiescent power for next-generation implants. IEEE J. Solid State Circuits **49**, 2812–2824 (2014)
16. H. Bhamra, J.W. Tsai, Y.W. Huang, Q. Yuan, P. Irazoqui, 21.3 a sub-mm^3 wireless implantable intraocular pressure monitor microsystem, in *2017 IEEE International Solid-State Circuits Conference (ISSCC)*, Feb 2017, pp. 356–357

M. Alhawari et al., *Energy Harvesting for Self-Powered Wearable Devices*,
Analog Circuits and Signal Processing, DOI 10.1007/978-3-319-62578-2

17. T.C. Chang, M.L. Wang, J. Charthad, M.J. Weber, A. Arbabian, 27.7 a 30.5mm^3 fully packaged implantable device with duplex ultrasonic data and power links achieving 95kb/s with $<10^{-4}$ BER at 8.5cm depth, in *2017 IEEE International Solid-State Circuits Conference (ISSCC)*, Feb 2017, pp. 460–461
18. F. Zhang, Y. Zhang, J. Silver, Y. Shakhsheer, M. Nagaraju, A. Klinefelter, J. Pandey, J. Boley, E. Carlson, A. Shrivastava, B. Otis, B. Calhoun, A batteryless 19μW MICS/ISM-band energy harvesting body area sensor node SoC, in *2012 IEEE International Solid-State Circuits Conference Digest of Technical Papers (ISSCC)* (2012), pp. 298–300
19. J. Cheng, L. Xia, C. Ma, Y. Lian, X. Xu, C.P. Yue, Z. Hong, P.Y. Chiang, A near-threshold, multi-node, wireless body area sensor network powered by RF energy harvesting, in *Proceedings of the IEEE 2012 Custom Integrated Circuits Conference*, Sept 2012, pp. 1–4
20. A. Roy, A. Klinefelter, F.B. Yahya, X. Chen, L.P. Gonzalez-Guerrero, C.J. Lukas, D.A. Kamakshi, J. Boley, K. Craig, M. Faisal, S. Oh, N.E. Roberts, Y. Shakhsheer, A. Shrivastava, D.P. Vasudevan, D.D. Wentzloff, B.H. Calhoun, A 6.45μW self-powered SoC with integrated energy-harvesting power management and ULP asymmetric radios for portable biomedical systems. IEEE Trans. Biomed. Circuits Syst. **9**, 862–874 (2015)
21. A. Usman, M. Mukhtar, R.M. Umer, Mems based wireless energy harvesting mechanism for ECG and BCG application, in *2014 International Conference on Emerging Technologies (ICET)*, Dec 2014, pp. 142–146
22. D. Bol, E.H. Boufouss, D. Flandre, J.D. Vos, A 0.48mm^2 5μW-10mW indoor/outdoor PV energy-harvesting management unit in a 65nm SoC based on a single bidirectional multi-gain/multi-mode switched-cap converter with supercap storage, in *European Solid-State Circuits Conference (ESSCIRC), ESSCIRC 2015 - 41st*, Sept 2015, pp. 241–244
23. X. Zhang, Z. Zhang, Y. Li, C. Liu, Y.X. Guo, Y. Lian, A 2.89 μW clockless wireless dry-electrode ECG SoC for wearable sensors, in *Solid-State Circuits Conference (A-SSCC), 2015 IEEE Asian*, Nov 2015, pp. 1–4
24. S. Izumi, K. Yamashita, M. Nakano, T. Nakagawa, Y. Kitahara, K. Yanagida, S. Yoshimoto, H. Kawaguchi, H. Kimura, K. Marumoto, T. Fuchikami, Y. Fujimori, H. Nakajima, T. Shiga, M. Yoshimoto, A normally-off ECG-SoC with noise tolerant heart rate extractor for wearable healthcare systems, in *2014 IEEE Biomedical Circuits and Systems Conference (BioCAS) Proceedings*, Oct 2014, pp. 280–283
25. F. Gonzalez, *Advanced and Post Lithium-Ion Batteries 2016–2026: Technologies, Markets, Forecasts* (IDTechEx, Cambridge, 2015)
26. H.S. Han, Y.H. Kim, S.Y. Kim, S. Um, J.M. Hyun, Performance measurement and analysis of a thermoelectric power generator, in *IEEE Intersociety Conference on Thermal and Thermomechanical Phenomena in Electronic Systems (ITherm)* (2010), pp. 1–7
27. Y. Ramadass, A. Chandrakasan, A battery-less thermoelectric energy harvesting interface circuit with 35 mV startup voltage. IEEE J. Solid State Circuits **46**(1), 333–341 (2011)
28. S. Lineykin, S. Ben-Yaakov, Modeling and analysis of thermoelectric modules. IEEE Trans. Ind. Appl. **43**(2), 505–512 (2007)
29. A.J. Minnich, M.S. Dresselhaus, Z.F. Ren, G. Chen, Bulk nanostructured thermoelectric materials: current research and future prospects. Energy Environ. Sci. **2**(5), 466–479 (2009) (Invited, peer-reviewed. Among the top 10 downloaded articles in EES for July, August, December 2009, January-May, July-December 2010, January 2011)
30. D. Dragoman, M. Dragoman, Giant thermoelectric effect in graphene. Appl. Phys. Lett. **91**, 203116 (2007). doi: 10.1063/1.2814080. http://aip.scitation.org/doi/full/10.1063/1.2814080
31. L. Mahmoud, Y. Abdul Samad, M. Alhawari, B. Mohammad, K. Liao, M. Ismail, Combination of PVA with graphene to improve the seebeck coefficient for thermoelectric generator applications. J. Electron. Mater. **44**(1), 420–424 (2015)
32. Y.-K. Teh, P. Mok, Design of transformer-based boost converter for high internal resistance energy harvesting sources with 21 mV self-startup voltage and 74% power efficiency. IEEE J. Solid State Circuits **49**(11), 2694–2704 (2014)
33. M. Wahbah, M. Alhawari, B. Mohammad, H. Saleh, M. Ismail, Characterization of human body-based thermal and vibration energy harvesting for wearable devices. IEEE J. Emerging Sel. Top. Circuits Syst. **4**, 354–363 (2014)

34. Tellurex, Thermoelectric specifications (2012), http://www.tellurex.com/, pp. 1–2
35. V. Leonov, T. Torfs, P. Fiorini, C. Van Hoof, Thermoelectric converters of human warmth for self-powered wireless sensor nodes. IEEE Sens. J. **7**, 650–657 (2007)
36. R. Vullers, R. Schaijk, H. Visser, J. Penders, C. Van Hoof, Energy harvesting for autonomous wireless sensor networks. IEEE Solid-State Circuits Mag. **2**, 29–38 (2010)
37. V. Leonov, R. Vullers, Thermoelectric generators on living beings, in *European Conference on Thermoelectrics* (2007)
38. M. Lossec, B. Multon, H. Ben Ahmed, C. Goupil, Thermoelectric generator placed on the human body: system modeling and energy conversion improvements. Eur. Phys. J. Appl. Phys. **52**, 11103 (2010)
39. A. Mirocha, P. Dziurdzia, Improved electrothermal model of the thermoelectric generator implemented in spice, in *ICSES'08. International Conference on Signals and Electronic Systems, 2008* (2008), pp. 317–320
40. V. Leonov, T. Torfs, P. Fiorini, C. Van Hoof, Thermoelectric converters of human warmth for self-powered wireless sensor nodes. IEEE J. Sens. **7**, 650–657 (2007)
41. C. Lu, V. Raghunathan, K. Roy, Efficient design of micro-scale energy harvesting systems. IEEE J. Emerging Sel. Top. Circuits Syst. **1**, 254–266 (2011)
42. S. Beeby, M. Tudor, N. White, Energy harvesting vibration sources for microsystems applications. Meas. Sci. Technol. **17**, R175–R195 (2006)
43. C. Williams, R. Yates, Analysis of a micro-electric generator for microsystems, in *International Conference on Solid-State Sensors Actuators Microsystems* (1995), pp. 369–372
44. N.E. DuToit, B L. Wardle, S.-G. Kim, Design considerations for MEMS-scale piezoelectric mechanical vibration energy harvesters. Integr. Ferroelectr. **71**, 121–160 (2005)
45. Y. Ramadass, Energy processing circuits for low-power applications. Ph.D. dissertation, Massachusetts Institute of Technology, 2009
46. MIDE and Volture, Piezoelectric energy harvesters (2013), pp. 1–24, http://www.mide.com/pdfs/Volture_Datasheet_001.pdf
47. MIDE and Volture, Energy harvesting electronics (2013), pp. 1–8, http://www.mide.com/pdfs/volture_EHE004_Datasheet.pdf
48. MIDE, Slam stick data logger datasheet (2010), pp. 1–15, http://www.mide.com/pdfs/slamstick_datasheet.pdf
49. Y. Ramadass, A. Chandrakasan, An efficient piezoelectric energy harvesting interface circuit using a bias-flip rectifier and shared inductor. IEEE J. Solid State Circuits **45**, 189–204 (2010)
50. J.L. Wardlaw, A.I. Karsilayan, Self-powered rectifier for energy harvesting applications. IEEE J. Emerging Sel. Top. Circuits Syst. **1**(3), 308–320 (2011)
51. M. Wahbah, B. Mohammad, Piezo electric energy harvester and its interface circuit: opportunities and challenges, in *IEEE International Conference on Electronics, Circuits, and Systems* (2013)
52. National Instruments, NI ELVIS II Series Specifications (2009), p. 11, http://www.ni.com/pdf/manuals/372590b.pdf
53. P. Mitcheson, T. Green, E. Yeatman, A. Holmes, Architectures for vibration-driven micropower generators. J. Microelectromech. Syst. **13**, 429–440 (2004)
54. S. Stanzione, C. Van Liempd, R. Van Schaijk, N. Yasuyuki, R. Yazicioglu, C. Van Hoof, A self-biased 5-to-60V input voltage and 25-to-1600μW integrated DC-DC buck converter with fully analog MPPT algorithm reaching up to 88% end-to-end efficiency, in *IEEE Solid-State Circuits Conference Digest of Technical Papers* (2013), pp. 74–76
55. D.G.A. Nimo, L.M. Reindl, Optimization of passive low power wireless electromagnetic energy harvesters. Sensors **12**, 13636–13663 (2012)
56. X. Lu, P. Wang, D. Niyato, D.I. Kim, Z. Han, Wireless networks with RF energy harvesting: a contemporary survey. IEEE Commun. Surv. Tutorials **17**, 757–789 (2015)
57. F.L.D. Bouchouicha, L. Ventura, Ambient RF Energy Harvesting, in *International Conference on Renewable Energies and Power Quality* (2010)
58. J.M. Barcak, H.P. Partal, Efficient RF energy harvesting by using multiband microstrip antenna arrays with multistage rectifiers, in *2012 IEEE Subthreshold Microelectronics Conference (SubVT)*, Oct 2012, pp. 1–3

59. J. Kwong, A. Chandrakasan, An energy-efficient biomedical signal processing platform. IEEE J. Solid State Circuits **46**, 1742–1753 (2011)
60. J. Hu, M. Ismail, *CMOS High Efficiency On-Chip Power Management* (Springer, New York, 2011)
61. D. Kilani, M. Alhawari, B. Mohammad, H. Saleh, M. Ismail, An efficient switched-capacitor dc-dc buck converter for self-powered wearable electronics. IEEE Trans. Circuits Syst. Regul. Pap. **63**, 1557–1566 (2016)
62. W. Jung, S. Oh, S. Bang, Y. Lee, D. Sylvester, D. Blaauw, 23.3 a 3nW fully integrated energy harvester based on self-oscillating switched-capacitor DC-DC converter, in *Solid-State Circuits Conference Digest of Technical Papers (ISSCC), 2014 IEEE International*, Feb 2014, pp. 398–399
63. C. Tse, S. Wong, M. Chow, The simplest lossless switched-capacitor AC/DC converter, in *Power Electronics Specialists Conference, PESC'94 Record., 25th Annual IEEE*, vol. 2, Jun 1994, pp. 1212–1217
64. J. Xiao, A. Peterchev, J. Zhang, S. Sanders, An ultra-low-power digitally-controlled buck converter IC for cellular phone applications, in *Applied Power Electronics Conference and Exposition, 2004. APEC'04. Nineteenth Annual IEEE*, vol. 1 (2004), pp. 383–391
65. T. Eichhorn, Boost Converter Efficiency Through Accurate Calculations (2008). http://www.powerelectronics.com/power-management/boost-converter-efficiency-through-accurate-calculations
66. S. Bandyopadhyay, P. Mercier, A. Lysaght, K. Stankovic, A. Chandrakasan, A 1.1nW energy harvesting system with 544pW quiescent power for next-generation implants, in *IEEE International Solid-State Circuits Conference Digest of Technical Papers (ISSCC)*, Feb 2014, pp. 396–397
67. R. Damodaran Prabha, G. Rincon-Mora, 0.18-μm light-harvesting battery-assisted charger-supply CMOS system. IEEE Trans. Power Electron. **31**, 2950–2958 (2016)
68. J. Kim, C. Kim, A DC-DC boost converter with variation-tolerant MPPT technique and efficient ZCS circuit for thermoelectric energy harvesting applications. IEEE Trans. Power Electron. **28**(8), 3827–3833 (2013)
69. J.-P. Im, S.-W. Wang, K.-H. Lee, Y.-J. Woo, Y.-S. Yuk, T.-H. Kong, S.-W. Hong, S.-T. Ryu, G.-H. Cho, A 40mV transformer-reuse self-startup boost converter with MPPT control for thermoelectric energy harvesting, in *2012 IEEE International Solid-State Circuits Conference Digest of Technical Papers (ISSCC)* (2012), pp. 104–106
70. T. Instrument (2012), http://www.ti.com/lit/ds/symlink/bq25504.pdf
71. H. Lhermet, C. Condemine, M. Plissonnier, R. Salot, P. Audebert, M. Rosset, Efficient power management circuit: from thermal energy harvesting to above-IC microbattery energy storage. IEEE J. Solid State Circuits **43**(1), 246–255 (2008)
72. P.-S. Weng, H.-Y. Tang, P.-C. Ku, L.-H. Lu, 50 mV-input batteryless boost converter for thermal energy harvesting. IEEE J. Solid State Circuits **48**(4), 1031–1041 (2013)
73. A. Blanco, G. Rincon-Mora, Bootstrapping and resetting CMOS starter for thermoelectric and photovoltaic chargers. IEEE Trans. Circuits Syst. Express Briefs **PP**(99), 1–1 (2017)
74. P.-H. Chen et al., Startup techniques for 95 mV step-up converter by capacitor pass-on scheme and tuned oscillator with fixed charge programming. IEEE J. Solid State Circuits **47**(5), 1252–1260 (2012)
75. E. Carlson, K. Strunz, B. Otis, A 20 mV input boost converter with efficient digital control for thermoelectric energy harvesting. IEEE J. Solid State Circuits **45**(4), 741–750 (2010)
76. P.-H. Chen, P.-Y. Fan, An 83.4% peak efficiency single-inductor multiple-output based adaptive gate biasing DC-DC converter for thermoelectric energy harvesting. IEEE Trans. Circuits Syst. Regul. Pap. **62**, 405–412 (2015)
77. I. Doms, P. Merken, C. Van Hoof, R. Mertens, Capacitive power management circuit for micropower thermoelectric generators with a 1.4 a controller. IEEE J. Solid State Circuits **44**(10), 2824–2833 (2009)

78. S. Bandyopadhyay, A. Chandrakasan, Platform architecture for solar, thermal, and vibration energy combining with MPPT and single inductor. IEEE J. Solid State Circuits **47**(9), 2199–2215 (2012)
79. Y. Ye, T. Liu, M. Chen, S. Nassif, Y. Cao, Statistical modeling and simulation of threshold variation under random dopant fluctuations and line-edge roughness. IEEE Trans. Very Large Scale Integr. VLSI Syst. **19**(6), 987–996 (2011)
80. S. Sleiman, A. Akour, W. Khalil, M. Ismail, Millimeter-wave BIST and BISC using a high-definition sub-ranged detector in 90nm CMOS, in *2010 53rd IEEE International Midwest Symposium on Circuits and Systems (MWSCAS)* (2010), pp. 477–480
81. M. Alhawari, B. Mohammad, H. Saleh, M. Ismail, An efficient zero current switching control for L-based DC-DC converters in TEG applications. IEEE Trans. Circuits Syst. Express Briefs **64**, 294–298 (2017)
82. S. Kim, G. Rincon-Mora, 23.4 dual-source single-inductor 0.18μm CMOS charger-supply with nested hysteretic and adaptive on-time PWM control, in *IEEE International Solid-State Circuits Conference Digest of Technical Papers (ISSCC)*, Feb 2014, pp. 400–401
83. P.-H. Chen, C.-S. Wu, K.-C. Lin, A 50nW-to-10mW output power tri-mode digital buck converter with self-tracking zero current detection for photovoltaic energy harvesting, in *Solid- State Circuits Conference - (ISSCC), 2015 IEEE International*, Feb 2015, pp. 1–3
84. M. Alhawari, B. Mohammad, H. Saleh, M. Ismail, An all-digital, CMOS zero current switching circuit for thermal energy harvesting, in *2015 European Conference on Circuit Theory and Design (ECCTD)*, Aug 2015, pp. 1–4
85. N. Bayasi, T. Tekeste, H. Saleh, B. Mohammad, A. Khandoker, M. Ismail, Low-power ECG-based processor for predicting ventricular arrhythmia. IEEE Trans. Very Large Scale Integr. VLSI Syst. **PP**(99), 1–13 (2015)
86. M. Pollak, L. Mateu, P. Spies, DC-DC-converter with input polarity detector for thermogenerators, in *International Conference on Micro and Nanotechnology for Power Generation and Energy Conversion Applications (PowerMEMS 2015)*, Dec 2009
87. Linear-Technology (LTC3109) (2012), http://www.linear.com
88. M. Alhawari, B. Mohammad, H. Saleh, M. Ismail, An efficient polarity detection technique for thermoelectric harvester in L-based converters. IEEE Trans. Circuits Syst. Regul. Pap. **64**, 705–716 (2017)
89. N. Mohan (ed.), *Power Electronics A First Course* (Wiley, New Haven, CT, 2012)
90. T. Na, S.-H. Woo, J. Kim, H. Jeong, S.-O. Jung, Comparative study of various latch-type sense amplifiers. IEEE Trans. Very Large Scale Integr. VLSI Syst. **22**, 425–429 (2014)
91. B. Mohammad, P. Dadabhoy, K. Lin, P. Bassett, Comparative study of current mode and voltage mode sense amplifier used for 28nm SRAM, in *2012 24th International Conference on Microelectronics (ICM)*, Dec 2012, pp. 1–6
92. Y. Ramadass, Energy harvesters and energy processing circuits. ISSCC tutorial, February 2013
93. M. Wahbah, M. Alhawari, B. Mohammad, H. Saleh, M. Ismail, Characterization of human body-based thermal and vibration energy harvesting for wearable devices. IEEE J. Emerging Sel. Top. Circuits Syst. **4**, 354–363 (2014)
94. M. Alhawari, B. Mohammad, H. Saleh, M. Elnaggar, An efficient zero current switching control for L-based DC-DC converters in TEG applications. IEEE Trans. Circuits Syst. Express Briefs **PP**(99), 1–1 (2016)
95. N.V. Desai, Y.K. Ramadass, A.P. Chandrakasan, A bipolar +-40 mV self-starting boost converter with transformer reuse for thermoelectric energy harvesting, in *2014 IEEE/ACM International Symposium on Low Power Electronics and Design (ISLPED)*, Aug 2014, pp. 221–226
96. A. Romani, A. Camarda, A. Baldazzi, M. Tartagni, A micropower energy harvesting circuit with piezoelectric transformer-based ultra-low voltage start-up, in *2015 IEEE/ACM International Symposium on Low Power Electronics and Design (ISLPED)*, July 2015, pp. 279–284

97. A. Shrivastava et al., A 10 mV-input boost converter with inductor peak current control and zero detection for thermoelectric and solar energy harvesting with 220 mV cold-start and −14.5 dbm, 915 MHz RF kick-start. IEEE J. Solid State Circuits **50**, 1820–1832 (2015)
98. N. Guilar, R. Amirtharajah, P. Hurst, S. Lewis, An energy-aware multiple-input power supply with charge recovery for energy harvesting applications, in *Solid-State Circuits Conference - Digest of Technical Papers, 2009. ISSCC 2009. IEEE International*, Feb 2009, pp. 298–299, 299a
99. A. Romani, M. Filippi, M. Tartagni, Micropower design of a fully autonomous energy harvesting circuit for arrays of piezoelectric transducers. IEEE Trans. Power Electron. **29**, 729–739 (2014)
100. J. Colomer-Farrarons, P. Miribel-Catala, A. Saiz-Vela, J. Samitier, A multiharvested self-powered system in a low-voltage low-power technology. IEEE Trans. Ind. Electron. **58**, 4250–4263 (2011)
101. T. Kang, S. Kim, C. Hyoung, S. Kang, K. Park, An energy combiner for a multi-input energy-harvesting system. IEEE Trans. Circuits Syst. Express Briefs **62**, 911–915 (2015)
102. Y.-K. Teh, P. Mok, A piezoelectric energy harvesting interface circuit using one-shot pulse transformer boost converter based on water bucket fountain strategy, in *2014 IEEE International Symposium on Circuits and Systems (ISCAS)*, June 2014, pp. 1993–1996
103. D. Carli, D. Brunelli, L. Benini, M. Ruggeri, An effective multi-source energy harvester for low power applications, in *Design, Automation Test in Europe Conference Exhibition (DATE), 2011*, March 2011, pp. 1–6
104. Y. Tan, S. Panda, Energy harvesting from hybrid indoor ambient light and thermal energy sources for enhanced performance of wireless sensor nodes. IEEE Trans. Ind. Electron. **58**, 4424–4435 (2011)
105. T. Tekeste, H. Saleh, B. Mohammad, A. Khandoker, M. Elnaggar, A nano-watt ecg feature extraction engine in 65nm technology. IEEE Trans. Circuits Syst. Express Briefs **PP**(99), 1–1 (2017)
106. M. Alhawari, T. Tekeste, B. Mohammad, H. Saleh, M. Ismail, Power management unit for multi-source energy harvesting in wearable electronics, in *2016 IEEE 59th International Midwest Symposium on Circuits and Systems (MWSCAS)*, Oct 2016, pp. 1–4
107. N. Bayasi, T. Tekeste, H. Saleh, B. Mohammad, A. Khandoker, M. Ismail, Low-power ECG-based processor for predicting ventricular arrhythmia, in IEEE Trans. Very Large Scale Integr. VLSI Syst. **PP**(99), 1–13 (2015)

Index

M. Alhawari et al., *Energy Harvesting for Self-Powered Wearable Devices*, Analog Circuits and Signal Processing, DOI 10.1007/978-3-319-62578-2

V

W

Z

Kolibri 360 GmbH
Ferdinand-Jühlke-Straße 7
99095 Erfurt, Deutschland
produktsicherheit@kolibri360.de